A Volume in The Laboratory Animal Pocket Reference Series

The Laboratory
SWINE

Second Edition

D0026481

The Laboratory Animal Pocket Reference Series

Series Editor
Mark A. Suckow, D.V.M.
Freimann Life Science Center
University of Notre Dame
South Bend, Indiana

Published Titles

The Laboratory Canine
The Laboratory Cat
The Laboratory Guinea Pig
The Laboratory Hamster and Gerbil
The Laboratory Mouse
The Laboratory Nonhuman Primate
The Laboratory Rabbit, Second Edition
The Laboratory Rat
The Laboratory Small Ruminant
The Laboratory Swine, Second Edition
The Laboratory Xenopus sp.

A Volume in The Laboratory Animal Pocket Reference Series

The Laboratory
SWINE

Second Edition

Peter J. A. Bollen

University of Southern Denmark
Odense, Denmark

Axel K. Hansen

University of Copenhagen
Denmark

Aage K. Olsen Alstrup

Aarhus University Hospital
Denmark

CRC Press
Taylor & Francis Group
Boca Raton London New York

CRC Press is an imprint of the
Taylor & Francis Group, an **informa** business

CRC Press
Taylor & Francis Group
6000 Broken Sound Parkway NW, Suite 300
Boca Raton, FL 33487-2742

© 2010 by Taylor and Francis Group, LLC
CRC Press is an imprint of Taylor & Francis Group, an Informa business

No claim to original U.S. Government works

Printed in the United States of America on acid-free paper
10 9 8 7 6 5 4 3 2 1

International Standard Book Number: 978-1-4398-1528-1 (Paperback)

Library of Congress Cataloging-in-Publication Data

Bollen, Peter J. A.
　　　The laboratory swine / Peter J.A. Bollen, Axel K. Hansen, and Aage Kristian Olsen Alstrup. -- 2nd ed.
　　　p. cm. -- (The laboratory animal pocket reference series)
　　Includes bibliographical references and index.
　　ISBN 978-1-4398-1528-1 (alk. paper)
　　1. Swine as laboratory animals--Handbooks, manuals, etc. I. Hansen, Axel Kornerup.
II. Alstrup, Aage Kristian Olsen. III. Title.

SF407.S97B66 2010
636.4'0885--dc22
　　　　　　　　　　　　　　　　　　　　　　　　　　　　　　　　　　　2009044299

Visit the Taylor & Francis Web site at
http://www.taylorandfrancis.com

and the CRC Press Web site at
http://www.crcpress.com

contents

preface

Laboratory swine have become important animal models, although the number of swine used in research stands no comparison with the number of rodents used. Since the first edition of *The Laboratory Swine* was published, a further increase in the number of swine has occurred, and swine are replacing other non-rodent species, such as dogs and primates, more and more in toxicity studies.

This book in the Laboratory Animal Pocket Reference Series was written to provide a source of information on laboratory swine. It is aimed at animal caretakers, technicians, and investigators; the laboratory animal veterinarian may also want to use this book as a reference handbook. This book provides an overview, but not an in-depth review, of the use of swine. Extensive references are given for further study.

The second edition has been updated, and especially the section on anesthesia, in Chapter 4, has been largely revised. It is our hope that this book will contribute to the humane use of laboratory swine, with the housing, husbandry, veterinary care, and experimental techniques that are appropriate to the species-specific physiology and behavior of swine.

The authors wish to thank Niels-Christian Ganderup, of Ellegaard Göttingen Minipigs, for providing the majority of the photographs for this book. Dr. Guy Bouchard of Sinclair Bio Resources is acknowledged for providing a photograph of Sinclair minipigs.

authors

Peter J.A. Bollen, Ph.D., is the head of the Biomedical Laboratory, the central animal facility of the University of Southern Denmark, Odense, where he also received his Ph.D. degree. He studied experimental zoology and laboratory animal science in Utrecht, the Netherlands, and laboratory animal science in London, England.

Axel Kornerup Hansen, Dr. Vet. Sci. DVM, is a professor of Laboratory Animal Science and Welfare at the Royal Veterinary and Agricultural University in Copenhagen, Denmark, from which he also graduated. He is the head of the Department of Veterinary Disease Biology, one of the four veterinary departments at the Faculty of Life Sciences of the University of Copenhagen, Denmark.

Aage Kristian Olsen Alstrup, Ph.D. DVM, is a graduate of the Royal Veterinary and Agricultural University in Copenhagen, Denmark, from which he also received his Ph.D. degree. He is responsible for animal models at the PET Center of Aarhus University Hospital, Denmark.

important biological features

Swine are important models in biomedical research. Their anatomical and physiological properties have many similarities to human anatomy and physiology, and many human disorders can be studied in swine. Swine have also gained importance as a non-rodent species in drug testing, especially for drugs that need to be administered orally or transdermally. There are many different swine breeds, varying enormously in size. Besides the breeds generally used in pork production, miniature breeds have been developed especially for laboratory use. In this book the term *swine* will refer to the large breeds, generally from an agricultural source, whereas the term *minipig* will be used for referring to the miniature breeds.

Since their development at the beginning of the 1950s, originally for the purpose of radiation biology, minipigs have been used in toxicity testing as a non-rodent species. The small size of minipigs is obviously advantageous for housing and handling, but also for the quantity of test compound, which usually is available in small amounts only.

breeds

The wild boar, *Sus scrofa*, is the ancestor of all modern breeds of swine. The first evidence of domestication in Europe is some 3,500 years old, although the cradle of the domesticated swine is claimed to have been in China about 10,000 years ago.[1] The American continent has no indigenous wild swine, aside from the distantly related Tayassuidae (peccaries). *Sus scrofa* is a member of the swine family (Suidae), which comprises species of non-ruminant, even-toed ungulates. Suidae are omnivores, and are divided into five genera with a total of 15 species (**Table 1**).

TABLE 1 CLASSIFICATION OF SUIFORMES

Order	Suborder	Family	Genus	Species	Common Name
Artiodactyla	Suiformes	Hippopotamidae		*Hippopotamus amphibius*	Hippopotamus
				Hexaprotodon liberiensis	Pigmy hippopotamus
		Tayassuidae		*Pecari tajacu*	Collared peccary
				Tayassu pecari	White-lipped peccary
				Catagonus wagneri	Chacoan peccary
		Suidae	*Babyrousa*	*Babyrousa babyrussa*	Hairy babirusa
				Babyrousa celebensis	Sulawesi babirusa
				Babyrousa togeanensis	Togian babirusa
			Phacochoerus	*Phacochoerus africanus*	Common warthog
				Phacochoerus aethiopicus	Somali warthog
			Potamochoerus	*Potamochoerus porcus*	Red river hog
				Potamochoerus larvatus	Bushpig
			Hylochoerus	*Hylochoerus meinertzhageni*	Giant forest hog
			Sus	*Sus scrofa*	Eurasian wild boar
				Sus salvanius	Pygmy hog
				Sus verrucosus	Javan warty pig
				Sus barbatus	Bearded pig
				Sus celebensis	Sulawesi warty pig
				Sus philippensis	Philippine warty pig
				Sus cebifrons	Visayan warty pig

Through selection, numerous breeds of swine have been developed. Existing breeds are relatively young in comparison to the history of the domestication of swine. Local Celtic swine with lop ears were widespread in European countries in the 18th and 19th centuries, and are considered to be the primogenitor of the Landrace. Crosses with Chinese breeds in the early 1800s, and with English breeds in the late 1800s, established the Danish Landrace, which was widely used for the derivation of other national Landrace breeds. The Yorkshire or Large White, a breed with pointed, standing ears, originated in England, and was exported to many countries around the world during the last half of the 19th century. The breed association for the American Yorkshire was established in 1893. Swine are numerous worldwide. China is the leading swine-producing country, with 680 million, followed by the United States, with 106 million swine produced in 2006. The 25 countries of the European Union produced 259 million swine in 2006.[2]

A short description of common breeds follows, based on descriptions from Porter (1993)[1] and Jones (1998).[3] The average weights are from Sambraus (1992).[4]

The Landrace

The Landrace swine has a long, white-colored body and lop ears. It has a long head with a slightly concave nose line. The Landrace breed is extremely variable in type, with its own distinct characteristics in different countries. The average weight of sows is 273 kg (601 lb), with boars averaging 312 kg (687 lb).

The Yorkshire, or Large White

The Yorkshire, or Large White, is a large, white breed of medium length. It has pointed, standing ears and a concave nose line (**Figure 1**). The sow is very fertile, with a litter size of 11, and weighs on average 280 kg (617 lb). The average weight of boars is 320 kg (705 lb).

The Duroc

The Duroc is a red swine with a large frame. The color varies from light to deep red. It has semi-lop ears. The Duroc is good natured. The average weight of sows is 300 kg (661 lb), and of boars 350 kg (771 lb).

Figure 1 A typical hybrid large swine with characteristics of both Landrace and Yorkshire breeds. (Photo courtesy of Polfoto, Copenhagen.)

Figure 2 The Hampshire is a compact breed and often used in hybrid crosses. (Photo courtesy of Polfoto, Copenhagen.)

The Hampshire

Hampshires are black with a white belt over the shoulder (**Figure 2**). The body is compact, and the legs relatively short. The ears are pointed and standing. The nose line is concave. The average weight of sows is 280 kg (617 lb), and of boars 320 kg (705 lb).

The Piétrain

The Piétrain is a short breed with a broad, deep body. The color is white with gray or black patches. It has short, pointed, standing ears. This breed is used mainly to improve meat quality in hybrid breeds. The Piétrain is very susceptible to stress. The average weight of sows is 266 kg (586 lb), and of boars 277 kg (610 lb).

Other Breeds

In 1990, the National Association of Swine Records had eight purebred swine on its registry. Along with the Landrace, Yorkshire, Hampshire, and Duroc (described above), the Berkshire, Chester White, Poland China and Spotted were registered.[1] Currently, the National Swine Registry keeps records of only four pure-bred swine: the Landrace, Yorkshire, Hampshire, and Duroc. These breeds are the basic stock for modern production herds, which are hybrid breeds.

Hybrid Breeds

Purebred swine are almost exclusively found in breeding herds. Modern production herds, however, utilize cross-breeding to produce hybrid grower swine. Hybrid breeds dominate the market nowadays; consequently, swine procured for the laboratory from an agricultural source will most likely be of mixed breed. Quadruple crosses, such as Yorkshire x Landrace (YL) sows with Duroc x Hampshire boars (DH), are common in modern pork farming. The demanded characteristics are litter size, milk production, and mothering ability for the sow, and muscle mass, growth traits, and leanness for the boar. At the laboratory, hybrid breeds may display various phenotypes, even when purchased from the same source. This can be recognized by the presence of both lop-eared individuals and individuals with pointed, standing ears or the presence of reddish brown or black spots on the nose or back. It is important to appreciate that phenotypical variation generally leads to larger variation in research results.

The genetic flexibility of Suidae has been demonstrated by an accidental hybrid cross between a male babirusa, or pig-deer, (*Babyrousa babyrussa*) and a Danish Landrace female, resulting in a litter of five hybrid pigs, born in 2006 at the Copenhagen Zoo. Three surviving offspring were euthanized in 2008 for scientific analysis. There was evidence of infertility, as commonly seen in interspecies hybrids.

Miniature Breeds

Minipigs are bred especially for research purposes. Many of the present breeds of minipigs have their origin in the Minnesota (Hormel) minipig, which was first bred in 1949 at the Hormel Institute in Austin, Minnesota. Minipigs derived from this population are the Göttingen minipig (1961) and the Sinclair minipig (1965). Other breeds are the closely related Hanford (1958) and Pitman-Moore (1969) miniature swine. A Mexican feral swine was introduced into the laboratory in 1960 and referred to as the Yucatan minipig. Later, a smaller subline of the Yucatan was selected, resulting in the Yucatan micropig. Several other breeds have been established, but are often available only to local research groups. The most widely available minipigs are the Göttingen, Yucatan, Hanford, and Sinclair minipigs.

There are two weight categories of minipigs: a 35–70 kg (77–154 lb), and a 70–90 kg (154–198 lb) adult weight category. To the lighter category belong the Göttingen minipig (**Figure 3A**), Sinclair minipig (**Figure 3B**), and Yucatan micropig, whereas the Hanford and Yucatan minipig belong to the heavier category.

The Vietnamese Potbelly Pig

This is a small Asian black breed with wrinkled skin, especially on the head (**Figure 4**). The ears are short and standing. The abdomen is round, almost touching the ground. The Vietnamese potbelly pig is very fertile but less suited for the laboratory because of its heavy anatomy.

The Ossabaw Island Hog

Swine of Ossabaw Island, off the coast of Georgia, are descendants of Spanish pigs brought to the New World over 400 years ago. They are small pigs with pointed, standing ears, a heavy coat, and long snout. Ossabaw Island hogs have remained a distinct and isolated population that reflects its Spanish heritage. They are commonly used as models of type 2 diabetes.

Nomenclature

Since many breeds of swine exist, each with its own phenotypical characteristics, in scientific publications it is essential to report which breed was used for a particular study. This facilitates comparison of experimental results between laboratories, and may explain unexpected differences. In commercial breeding, breeds generally are referred to with

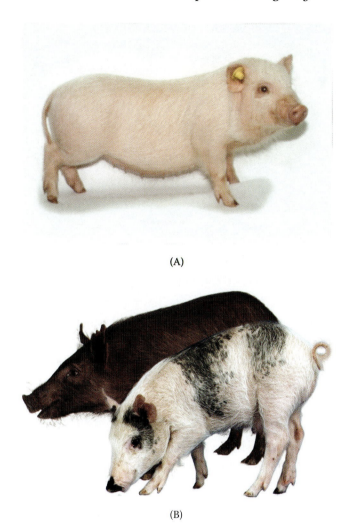

(A)

(B)

Figure 3 (A) The Göttingen minipig is a small breed with an adult weight of 35–45 kg. Here is a young animal with an age of 6 months and a weight of 12 kg. (Photo courtesy of Ellegaard Göttingen Minipigs, Soroe Landevej, Dalmose.) **(B)** The Sinclair S-1 miniature swine is a small breed in direct lineage with the Minnesota minipig. (Photo courtesy of Guy F. Bouchard, DVM, Sinclair Bio Resources, LLC, Auxvasse, Missouri.)

Figure 4 The Vietnamese potbelly swine is very fertile, but less suited for the laboratory. (Photo courtesy of Polfoto, Copenhagen.)

capitals, but this may not be desirable for the scientific community because of unfamiliarity with this system. A nomenclature system, as used for rodents, is not known for laboratory swine. The scientifically correct way of presenting data includes breed, gender, age, and weight, along with supplier and housing conditions.[5]

The definitions of common terms used for swine are given below:

swine, hog, pig	denotation for the species *Sus scrofa*; the first two terms are commonly used in American literature, whereas the third is used mainly in British literature
sow	sexually mature female
gilt	sexually immature female
boar	sexually mature male
barrow	castrated male
piglet	juvenile swine
porcine	adjective used in relation to swine

behavior

An important reason swine became the subject of domestication is their behavior. Their feeding, sexual, and social behaviors, as well as their typically docile nature and general adaptability, favored domestication. Compared to wild swine, domesticated swine are calmer, quieter, and less active, although they still share many behavioral characteristics with wild swine, such as social, feeding, and explorative behavior.[6]

Wild swine live in small social groups formed by sows and their offspring. Young males live in bachelor groups, whereas adult boars tend to be solitary. In groups of females, a social dominance order is quickly established. Generally the level of aggression, expressed mainly by butting and by biting the neck and ears, soon subsides in stable social groups. Sexually mature males may fight fiercely, especially in the presence of females.

Newborn swine establish a social order, the teat order, within a few days after birth. As soon as a teat order has been established, each piglet consistently suckles the same teat. Dominant piglets generally occupy the more productive anterior teats. If young swine from different litters are mixed after weaning, a new dominance hierarchy will be established. Ranking is the result of aggressive interaction, but may take place without overt aggression. It exists as long as a group is together. Subordinates that have been separated from a group will be attacked after reintroduction, whereas a dominant animal may be separated and reintroduced without any trouble.[7]

Wild swine spend most of the day rooting and collecting food. Their feeding and explorative behaviors are closely linked. During the day, swine eat many small portions. Swine are omnivorous and diurnal, with elevated activity during the evening. In the laboratory, activity of swine is related to the presence and activity of humans rather than to the light-dark cycle.[8]

anatomical and physiological features[9-13]

Integument and Skeleton

The skin of swine has anatomical and physical similarities with human skin; consequently, swine are often used as a model for percutaneous absorption, skin toxicology, and wound-healing studies.

Swine have short, sturdy bones. The number of ribs varies greatly, from 13 to 17, but most commonly 14 or 15 ribs are present. The bony structure of the thorax is considerably smaller than the external dimensions suggest. The skull is extremely sturdy, and has a largely extended frontal sinus. Swine have been used in osteoporosis research and cartilage-repair studies, but because of a high bone density and relatively short bones, swine are not the preferred species for orthopedic research.

Digestive System

Swine have an extensive dentition, with 44 elements. For this reason, they are often used in dental research. The formula for the permanent teeth is i3/3, c1/1, p4/4, m3/3. The canine teeth, or tusks, can be found in both boars and sows, but are most developed in boars. The tusks of boars grow throughout the animal's life. Swine are born with 8 teeth, the needle teeth (i1/1, c1/1). After two months, two more deciduous incisors and three of the deciduous premolars have erupted (i3/3, c1/1, p3/3). At 6 months, the first permanent teeth erupt. The swine is at least 18 months old before all permanent teeth have erupted.

Swine are monogastric, and the stomach has a large fundus and a diventriculum. The area where the esophagus enters the stomach has non-glandular mucosa and is prone to peptic ulcers. The bile duct and pancreatic duct enter the duodenum. Similar to humans, swine have an unbranched pancreatic duct, which makes swine a good model for studying pancreatic secretion. The jejunum and ileum contain numerous patches of lymph nodes. The large intestine is coiled and voluminous. Because of the structure and large volume, the cecum lies to the left.

Abdominal Organs

The liver has much fibrous tissue, giving the surface a net-like appearance. The pancreas has two lobes, which are difficult to identify. The kidneys have multiple papillae protruding into a central renal pelvis via the calyces. In structure and size they are similar to human kidneys. Lymph nodes are located along all major arteries in the abdomen. The uterine horns in sows are very long, and the cervix is tightly closed with mucosal ridges. The boar has large testes and various accessory genital glands. The ejaculate is voluminous, from 200 to 500 milliliters. The penis is fibrous, and has a corkscrewed tip. The preputial diverticulum produces a strong odor.

Cardiovascular System

The heart of swine is small in relation to body size. The coronary distribution is similar to that of humans, but the left azygous vein ends directly in the heart instead of in the caval vein. Swine develop atherosclerosis of the major arteries after ingestion of a lipidous diet over a relatively short period.

Pulmonary System

Swine have a tracheal bronchus to ventilate the right cranial lobe; therefore, endotracheal intubation should not continue as far as the tracheal bifurcation.

Lymphatic and Endocrine System

The lymph nodes of swine are inverted, with the follicular centers in the medulla. The parathyroids are not located near the thyroids, but are related to the thymus. The right adrenal lies in close contact with the caudal caval vein, which precludes total adrenalectomy.

normative values

Biological, hematological, clinical chemical, and respiratory and cardiovascular parameters are presented in **Tables 2** through **6**. When comparing data, it is important to compare data from age-matched rather that weight-matched swine.[14] However, variation between individuals and breeds may occur, due to genotype, environment, and experimental procedures.

TABLE 2 BASIC BIOLOGICAL PARAMETERS (RANGE) OF SWINE AND MINIATURE SWINE WITH ADULT WEIGHTS OF 35–70 KG AND 70–90 KG

		Miniature Swine	
Parameter	Swine	35–70 kg[a]	70–90 kg[b]
lifespan (yrs)	10–15	10–15	10–15
body temperature (°C)	38–39	37–38	38–39
chromosomes (2n)[15]	38	38	38
weight at birth (kg)	1.3–1.9	0.4–0.7	0.6–1.0
weight at 6 months (kg)	90–110	12–22	25–40
weight at 1 yr (kg)	150–180	25–40	45–70
weight at 2 yrs (kg)	200–300	35–55	70–90
energy intake[c] (MJ/day)[16]	49–56	13.8–21.9	21.9–29.2
food intake[c,d] (kg/day)	3.6–4.1	1.0–1.6	1.6–2.1
water intake (mL/kg/day)[17]	80–120	80–120	80–120

[a] Yucatan micropig, Göttingen and Sinclair minipigs
[b] Yucatan and Hanford minipigs
[c] Estimated by ME = $1.2 \times BW^{0.75}$ (MJ/day), calculated for a body weight (BW) at 6 months. ME = metabolizable energy requirement
[d] At voluntary food intake (*ad libitum*) with a diet with an energy concentration of 13.6 MJ/kg (3265 kcal/kg)

TABLE 3 HEMATOLOGICAL PARAMETERS (RANGE) OF SWINE AND MINIATURE SWINE WITH ADULT WEIGHTS OF 35–70 KG AND 70–90 KG

Parameter	Swine[13,18,19]	Miniature Swine	
		35–70 kg[a 13,20,21]	70–90 kg[b 13,22]
RBC (10^6/µL)	4.4–8.6	5.30–9.25	5.6–8.8
Hemoglobin (g/dL)	9–16.2	9–15.8	13.1–17
Hematocrit (%)	33.9–45.9	32–61	36.3–53.7
MCV (fL)	17.6–79.6	40–73	58.2–72.5
MCH (pg)	15.2–56.3	15.2–26.4	18.9–24.3
MCHC (%)	29.4–35.9	29.4–37.9	31.1–34.5
WBC (10^3/µL)	6.3–21.1	4.4–26.4	6.9–21.2
Segmented neutrophils (%)	22.0–60.6	10–80	18–94
Band neutrophils (%)	0–4.2	0–4	0–2
Lymphocytes (%)	38.1–73.1	14–87	21–71
Monocytes (%)	0–15	0–13	2.0–15
Eosinophils (%)	0–7.7	0–8	0–13
Basophils (%)	0–1.3	0–3	0–5
Platelets (10^3/µl)	220–665	148–898	217–770

[a] Yucatan micropig, Göttingen and Sinclair minipigs
[b] Yucatan and Hanford minipigs

TABLE 4 CLINICAL CHEMICAL PARAMETERS (RANGE) OF SWINE, AND MINIATURE SWINE WITH ADULT WEIGHTS OF 35–70 KG AND 70–90 KG

Parameter	Swine[13,24]	Miniature Swine	
		35–70 kg[a 13,20,21]	70–90 kg[b 22,23]
Glucose (mg/dL)	48–135	43–133	56–153
Creatinine (mg/dL)	1.2–2.0	0.5–1.6	1.2–2.0
Bilirubin (mg/dL)	0.0–0.3	0.0–0.2	0.0–0.3
Cholesterol (mg/dL)	50–140	39–131	47–173
Total protein (g/dL)	2.25–8.15	6.0–8.8	6.3–9.4
Albumin (g/dL)	0.5–4.3	2.9–3.8	4.1–5.6
Globulin (g/dL)	1.4–3.6	1.5–5.2	1.4–3.6
A:G ratio	1.1–3.5	0.9–1.7	1.1–3.5
Sodium (mEq/L)	133–153	132–146	142–153
Potassium (mEq/L)	3.1–6.2	3.5–7.4	3.9–5.2
Chloride (mEq/L)	96–117	94–140	95–114
Calcium (mg/dL)	5.5–15.7	8.6–12.6	9.3–11.6
Phosphorus (mg/dL)	4.8–9.8	4.9–9.8	5.0–8.3
AST (IU/L)	14–56	13–47	15–53
ALT (IU/L)	5–78	40–106	20–48
CK (IU/L)	52–326	105–6000	37–270
GGT (IU/L)	14–34	25–78	41–86
LDH (IU/L)	140–1155	462–1800	389–727

[a] Yucatan micropig, Göttingen and Sinclair minipigs
[b] Yucatan and Hanford minipigs

TABLE 5 RESPIRATORY AND CARDIOVASCULAR FUNCTION (MEAN ± SD) OF CONSCIOUS SWINE, AND MINIATURE SWINE WITH ADULT WEIGHTS OF 35–70 KG AND 70–90 KG

Parameter	Swine[13,25]	Miniature Swine	
		35–70 kg[a 13,26]	*70–90 kg*[b 13,14,27]
heart rate	105 ± 10.6	83 ± 15	105 ± 7
mean arterial blood pressure (mmHg)	102 ± 9.3	97 ± 14	89 ± 3
respiration rate	20 ± 2.9	20 ± 9	25 ± 4
arterial pH	7.48 ± 0.03	7.43 ± 0.03	7.48 ± 0.04
arterial pCO_2 (mmHg)	40 ± 2.3	40 ± 3	43 ± 4
arterial pO_2 (mmHg)	71 ± 3	109 ± 17	84 ± 4

[a] Yucatan micropig, Göttingen and Sinclair minipigs
[b] Yucatan and Hanford minipigs

TABLE 6 REPRODUCTION PARAMETERS OF SWINE, AND MINIATURE SWINE WITH ADULT WEIGHTS OF 35–70 KG AND 70–90 KG

Parameter	Swine[7]	Miniature Swine	
		35–70 kg[a 28–30]	*70–90 kg*[b 31]
sexual maturity (months)	6	4–5	6
minimum breeding age (months)	7	5–6	7
estrus cycle (days)	14	14	14
length of estrus (days)	3	3	3
length of pregnancy (days)	114	114	114
litter size	10–14	5–8	5–8
weight at birth (kg)	1.30	0.45–0.60	0.60–1.00
age at weaning (days)	28–35	28–35	28–35

[a] Yucatan micropig, Göttingen and Sinclair minipigs
[b] Yucatan and Hanford minipigs

husbandry

An important factor for the successful use of swine in biomedical research is knowledge about the specific husbandry requirements of swine. Housing, feeding, and care should be appropriate to the anatomy, physiology, and behavioral characteristics of this species. Changes in environmental conditions, feeding, and care may have an impact on experimental results in swine studies.[32]

housing

At a typical commercial breeding establishment, separate quarters are provided for

- Young stock from weaning to breeding age (growing quarters)
- Dry sows and gilts, and breeding boars (mating quarters)
- Pregnant sows (waiting quarters)
- Parturient and nursing sows (farrowing quarters)

In the growing quarters, swine are housed in groups of similar sex. Groups will live in good harmony when enough space per animal is available so that animals can move around freely to avoid aggressive interaction. Large groups or groups in a confined space will be unable to maintain a stable dominance order, resulting in a higher level of aggression.[33] Stable social groups are best established after weaning.

Young gilts and pregnant sows are group housed. In the farrowing quarters, sows are caged in special farrowing crates equipped with bars to prevent them from lying on their young. Breeding boars are housed individually.[7]

Figure 5 Group pen at a breeding facility with a solid concrete floor with epoxy coating. Cage furniture is provided for environmental enrichment of this easy-to-sanitize but barren environment. (Photo courtesy of Ellegaard Göttingen Minipigs, Soroe Landevej, Dalmose.)

Pigs kept at the laboratory can be housed individually or in small groups in floor pens. The floor can be made of solid concrete or raised grids. Concrete floors should have a rough surface for secure footing, and may be coated with an epoxy layer. Bedding, such as wood shavings or straw, should be provided. This gives good rooting and nesting possibilities for the animals, but is more laborious for sanitation. Grid floors provide good sanitation, since usually no bedding is provided. Grid floors can be made of galvanized steel, plastic-coated steel, or fiberglass, and may consist of parallel bars or have a mesh structure. Galvanized steel is least suitable since it conducts heat too well, whereas plastic-coated steel and fiberglass preserve heat more satisfactorily. However, room temperature is more critical when using raised grid floors compared to concrete floors with bedding, since the animals have no possibility of nesting and creating an insulated microenvironment. Attention should be paid to the spacing of the grids, so that hooves will not be damaged. When using parallel bars, a bar width of approximately 10 mm and a spacing of approximately 12 mm is appropriate for most sizes of swine. Plastic-coated, diamond-shaped grids (Tenderfoot®) with a spacing of 10–15 mm are a very satisfactory floor type for swine. When swine are housed on grid floors, hooves need to be trimmed at regular intervals, since hooves do not wear down on grid floors as they do on concrete floors (**Figure 5**).

Pens should be robust, since swine are forceful animals with a strong snout. They also rub their sides along the sides of the pen, pushing with considerable force. Concrete or brick walls are most often used, in combination with fencing of galvanized or stainless steel. No sharp edges or protruding bolts should be present. Especially when animals are single housed, fencing should be placed in such a way as to make it possible for the animals to see, smell, and hear each other. Swine are social animals, and need to have visual, olfactorial, and auditory contact if physical contact is not possible.

Young swine and miniature swine can be housed in dog facilities without major adaptations to the pens. Usually, lowering automated watering valves and feeding bowls is all that is necessary. Automated watering valves are preferred to water bowls, since swine drink considerable amounts of water, and water should be constantly available. Feeding bowls should be secured to the cage, and be made of a smooth material that is easy to clean. With group-housed swine, troughs are commonly used. Attention should be paid to competition. There should be enough space at the trough for each animal. Nevertheless, differences in feed uptake may occur when feeding from troughs, with a differentiation in body-weight development as a result. Hence, trough feeding is less suitable for the laboratory.

Outdoor housing is unsuitable in a research setting because of uncontrolled environmental conditions and risk of infection. In regions with mild climatic conditions, raising pigs in huts or kennels on pastures is gaining in popularity. For swine obtained from agricultural sources, inquiries must be made about housing type, since outdoor swine without exception need to be treated with anthelminthics when arriving at the laboratory.

Space guidelines for swine are described by the National Research Council[34] and the European Council Directive.[35] Since no consensus exists among the guidelines, common sense has to be applied when housing swine. The animals should be "provided with housing, environment, degree of movement, food, water and care which are appropriate to their health and well-being. Restrictions to the physiological and behavioral needs of experimental animals should be limited to the absolute minimum."[35] In general, less space is given to swine in agriculture settings than what is recommended for research swine. Space recommendations for research swine are given in **Table 7**.

environmental conditions

Elements of environmental conditions are temperature and humidity, ventilation, illumination, and noise.

TABLE 7 SPACE RECOMMENDATIONS FOR SWINE[34]

Number of Swine per Enclosure	Weight (kg)	Floor Area per Animal (ft²)	Floor Area per Animal (m²)
1	<15	8	0.72
	15–25	12	1.08
	25–50	15	1.35
	50–100	24	2.16
	100–200	48	4.32
	>200	>60	>5.40
2–5	<25	6	0.54
	25–50	10	0.90
	50–100	20	1.80
	100–200	40	3.60
	>200	>52	>4.68
>5	<25	6	0.54
	25–50	9	0.81
	50–100	18	1.62
	100–200	36	3.24
	>200	>48	>4.32

Temperature and Humidity

Since swine have relatively sparse hair and lack sweat glands, they are sensitive to temperature fluctuations and extreme temperatures. Especially when swine are housed on grid floors without bedding, flawless temperature control of the facility is necessary. Optimum environmental temperatures for swine vary with age. Newborn swine thrive best at temperatures of 30–38°C (86–100°F). This is achieved by partial floor heating in the farrowing pen, with the addition of a thermostatically controlled heat lamp.[7]

Swine are best housed at the thermoneutral temperature. At this temperature, relative humidity appears to be unimportant. Energy conversion is not influenced by the relative humidity at the thermoneutral temperature, which was found to be 29°C (84°F) in juvenile (6–8 weeks old), 24°C (75°F) in young (14–16 weeks old) and 17.4°C (63°F) in adult (34–36 weeks old) miniature swine.[36] These figures apply to swine generally, but swine housed on raised-grid floors without bedding should not have temperatures lower than 20°C (68°F) because of an expected higher heat loss. The relative humidity should be maintained at 50–70%.

Ventilation

Ventilation is essential to keep the concentration of potentially harmful gases low, especially in densely stocked quarters. Concentrations of NH_3

and H_2S should not exceed 10 ppm and 5 ppm respectively, and CO_2 levels should not rise beyond 0.15 vol-%. However, air circulation should not exceed 0.2 to 0.3 m/s for adult animals and 0.1 m/s for piglets.[7]

Generally, a ventilation rate of 10–15 changes per hour with fresh air is recommended, but the dimension of the room and the number of animals present need to be considered, especially to prevent over-ventilation, since swine are sensitive to draft. On the other hand, under-ventilation will lead to a raised environmental temperature and increased levels of noxious gases.

Illumination

A light-dark cycle of 12 hours is usually applied to swine, with the lights on from 6 a.m. to 6 p.m. A light intensity of 100 lux is appropriate, but in the mating quarters of breeding facilities, light intensity should be 200 lux, with the lights on from 6 a.m. to 8 p.m.

Noise

Swine are relatively insensitive to noise. As a matter of fact, they are quite noisy themselves, and staff are recommended to wear hearing protection when handling swine. However, sudden, explosive noises should be avoided, since these induce fright.

environmental enrichment

Swine are lively animals. Although they spend 70 to 80 percent of their time lying or sleeping, during the remaining time they are actively strolling around, exploring and rooting. When given the space, swine are clean in their habits, choosing specific sites for defecation and urination while keeping their sleeping area dry. The availability of clean, dry straw contributes substantially to the well-being of swine. Straw provides comfortable bedding and keeps the animals occupied with rooting and chewing activities. Unless experimental conditions require otherwise, the best way of accommodating swine and miniature swine is by housing them in groups of up to 10 or 15 animals in spacious pens with straw-covered floors.

Group Housing

Social grouping and the establishment of a social hierarchy are behavioral traits typical for swine. These behavioral patterns should be considered when housing swine. If possible, individual housing of swine should be avoided. With group-housed swine, uniformity of the group must be observed and trough space must be sufficient to permit all animals to feed at the same time. Ideally, partitions should be provided, enabling weaker individuals to avoid the more aggressive ones. Group size should not exceed 10 to 15 animals; otherwise, no lasting hierarchy will be established. If possible, groups should be kept together. When the animals have to be regrouped, the extent of fighting may be minimized by bringing unfamiliar animals together just before feeding or sleeping time, preferably in a newly cleaned pen. The individual scent of all animals of the group may be camouflaged with a strong fragrance (mint oil, cresol), and in critical cases animals could be tranquilized before being brought together.

Adult boars are solitary, and individual housing is appropriate. Nonetheless, it is possible to group house boars, with the exception of particularly aggressive individuals, provided they have been reared together or have been given the opportunity to get accustomed to each other under unconfined conditions. Barrows can be group housed under conditions similar to those of females.

Individual Housing

At the laboratory, individual housing is common. As described before, individually housed animals should have visual, olfactorial, and auditory contact with other swine in order to prevent them from being socially deprived. To a large extent, swine become attached to humans, especially when housed individually,[37,38] and daily positive contact with staff should be included in the animal care program.[39]

Most important to swine is the provision of items that can be manipulated, especially when no bedding is provided, to satisfy the needs of chewing and rooting. The following items can be considered elements of environmental enrichment:

- a chain, suspended from the side of the pen
- a hay-rack with straw or hay
- a nylon brush or broom head, attached to fencing
- a heavy ball, plastic crate, or plastic turf mat placed on the floor

A chain, a rope, and straw or hay in a hay-rack primarily satisfy the need for chewing, and they also provide material for manipulation with the snout. A brush or broom head, attached firmly to the fencing, is used for rubbing against and for biting. Balls, crates or turf mats are used for satisfying the need for rooting. For all elements of environmental enrichment it is essential to allow a good level of sanitation. The animals quickly lose interest if the items get soiled, especially items placed on the floor of the pen. To avoid soiling and to preserve the novelty of the items, balls, crates, or turf mats should be given to the animals only for a limited period daily.

nutrition

Most publications relating to swine nutrition focus on optimal utilization of nutrients with an eye on efficient growth, since feed accounts for two thirds of the costs of producing market-weight swine.[40]

Although laboratory swine nutrition is basically similar to production swine feeding, it may be desirable to control weight gain. Therefore, restricted feeding is often applied at the laboratory. Moreover, miniature swine are prone to obesity, so restricted feeding is necessary to maintain normal physiological conditions in them. There are indications that nutrient requirements of miniature swine differ from those of other swine,[41] but until further evidence has been produced, nutrient requirements of swine should be used as a basis for miniature swine nutrition.[42,43]

Nutrient Requirements

The National Research Council (NRC) provides detailed information on the composition of swine diets, including energy, protein and amino acids, minerals and vitamins, and diet formulation. However, the given nutrient requirements are minimum standards when feeding *ad libitum*, and should not be considered recommended allowances.[17] **Table 8** gives an overview of the nutrient requirements of swine with body weights from 5 to 80 kg.

Swine are able to utilize high-fiber diets because of fermentation in the cecum and large intestine. However, fiber levels over 7–10% inhibit growth. Dietary fiber also has an effect on the passage of food through the intestines, and diets with a high fiber content prolong the gastrointestinal transit time. Fiber levels of up to 15% have no effect, but levels exceeding 15% will increase the transit time. Especially when roughage, such as straw or hay, is fed, gastric emptying and intestinal transit time are significantly prolonged.[44] Low-energy diets,

TABLE 8 NUTRIENT REQUIREMENTS OF SWINE WITH
BODY WEIGHTS FROM 5 TO 80 KG[17]

Body Weight (kg)	ME[a] Content of Diet (MJ/kg)	Expected Feed Intake (g/day)	Crude Protein (%)
5–10	13.6	500	23.7
10–20	13.6	1,000	20.9
20–50	13.6	1,850	18.0
50–80	13.6	2,575	15.5

[a] Metabolic energy; 1 MJ = 240 kcal (13.6 MJ/kg = 3,265 kcal/kg)

especially meant as maintenance diets for miniature swine, usually have a relatively high fiber content of 13–15%.

The minimum level of minerals and vitamins is well defined,[17] and commercial diet producers may be expected to have included sufficient mineral and vitamin premix to cover the requirements of swine. However, when comparing levels from catalogue values and analysis reports, disturbing discrepancies may be found within and between brands.[41]

Feeding Levels

The general practice in swine feeding is *ad libitum* feeding. With this, modern production swine grow 2–3 kg (4.4–6.6 lb) per week, so that they weigh 80–100 kg (175–225 lb) at the age of 5–7 months, when generally they are slaughtered. This enormous weight increase may be undesirable at the laboratory. Therefore, restricted feeding is often applied at research facilities. In miniature swine, restricted feeding is essential, since they easily become obese when fed without restriction. From studies in rodents, it has become clear that a feeding level of 60% of the *ad libitum* food intake is sufficient to prevent obesity, fulfills nutrient needs, and leads to a longer, healthier life as compared to *ad libitum* feeding.[45] This may be applied to laboratory swine as well, but, more often, restricted feeding with a low-energy diet is applied to allow the animals a satisfying food volume intake. Whereas a standard swine diet has a metabolic energy content of 13.6 MJ/kg (3265 kcal/kg), a typical low-energy diet has a metabolic energy content of 9.5 MJ/kg (2275 kcal/kg). Feeding levels can be found in **Table 9**, and growth rates in **Figure 6**.

Clean drinking water should be available at all times. Therefore, automated watering systems (nipple or bowl type) are recommended.

TABLE 9 FEEDING LEVELS

Body Weight (BW in kg)	Expected Energy Intake (MJ/day)[a]	Daily Feeding (g/day)[b]	Restricted Feeding (g/day)[c]	Low Energy Feeding (g/day)[d]
5	4.00	290	175	255
10	6.75	500	300	425
20	11.35	835	500	715
50	22.55	1,660	995	1,425
80	32.10	2,360	1,415	2,025

[a] $ME = 1.2\ BW^{0.75}\ (MJ/day)$[16]
[b] metabolizable energy density of diet = 13.6 MJ/kg (3,265 kcal/kg)
[c] 60% of expected energy intake with a diet of ME=13.6 MJ/kg (3,265 kcal/kg)
[d] 60% of expected energy intake with a diet of ME= 9.5 MJ/kg (2,275 kcal/kg)

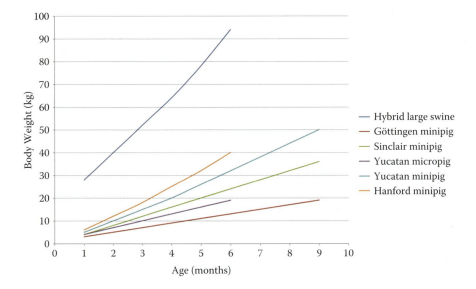

Figure 6 Weight development of different swine breeds depends on the level of feeding. Generally, large swine are fed *ad libitum* for securing rapid growth, whereas miniature breeds are fed restrictedly. (Data from http://www.minipigs.com/ and http://www.sinclairbioresources.com/.)

Although water is essential, surprisingly little research has focused on the water requirements of swine. True water consumption is often overestimated, since swine like to play with water, and wastage is often not taken into account.[17] Approximately 120 mL/kg water is used by growing swine fed a dry pelleted diet.[17,46] The delivery rate of water via automated watering devices should be about 0.5 L/min.

If automated watering devices are not available, the usual practice is to mix water with the diet, and then refill the trough with water after the swine have finished eating.

sanitation

Sanitation is important in maintaining the health of swine. Excreta and food remains are a substrate for microbial growth and attract insects and rodents that could act as vectors for diseases. Therefore, pens should be cleaned frequently, but sanitation should be carried out with minimal use of disinfectants to prevent chemical contamination of the animal environment.[47]

Frequency

Daily removal of excreta is good practice. When swine are housed on solid floors with bedding, removal of soiled bedding is essential. Soiled bedding represents the largest volume of waste material and could lead to capacity problems when disposing waste in bags or containers.

Swine housed on grid floors usually press feces through the grids with their hooves, while urine runs through. Under the grid floor, a sewer canal may be present through which the excreta can be flushed away. When large numbers of animals are present, animal rooms for swine should be constructed with a manure cellar under the grid floors. Here, excreta accumulate and are pumped to a large basin outside the facility at regular intervals. Grid floors should be checked daily, and feces accumulated in corners should be scraped away.

Pens should be washed weekly, although pens with grid floors are usually washed at longer intervals to prevent excessive amounts of liquid manure. A mild disinfectant may be used. Disinfectants should be rinsed away and solid floors swabbed dry before allowing the animals to return to their pen.

Monthly, pens should be sanitized thoroughly. High-pressure cleaning is commonly used in swine facilities. High-pressure cleaning should be carried out without additives to the water, because of aerosol formation. A detergent or disinfectant is usually added to low-pressure water to rinse with after high-pressure cleaning. Detergents or disinfectants should be rinsed away and solid floors swabbed dry before allowing the animals to return to their pens.

Methods

Sweeping should be minimized to prevent dust formation. Floors of pens, corridors, and utility rooms should be washed and swabbed dry after gross debris has been removed. Staff working with soiled bedding, especially wood shavings, should wear dust masks for protection.

High-pressure cleaning should include floors, walls, fencing, bowls and troughs, drinking devices, and items for environmental enrichment. Detergents or disinfectants should not be used before high-pressure cleaning, nor be added to high-pressure water, because of the risk of exposure to aerosols during high-pressure cleaning. Staff should wear protective clothing (watertight suit or apron), fluid shield masks, and hearing protection, and no animals should be present when using high-pressure cleaning.

A wide range of detergents and disinfectants are available, and suitable detergents and disinfectants should be chosen based on effectiveness and experience.[48,49]

---

transportation

Transportation is stressful for swine, and gentle handling is necessary. Since swine are prone to travel sickness, animals should be deprived of food 12 hours prior to and during transportation. If animals are transported over long periods, breaks should be included at least every 6 hours to offer the animals water. Hot, humid conditions during transportation should be avoided, as well as low temperatures and drafts. Ideally, acclimatized vehicles should be used to maintain environmental conditions similar to those at the animal facility.

Swine are generally transported freely standing in a truck or trailer. A space of 0.5 m²/100 kg (2.5 ft²/100 lb) body weight should be allowed. Floors must not be slippery; a slatted floor covered with wood shavings is commonly used. Large loading surfaces should be avoided, or be subdivided by partitions.

Young swine and miniature swine can be transported in portable dog kennels (such as Vari Kennel®, shown in **Figure 7**) or custom-made crates. Floors should be covered with wood shavings. Straw is not as suitable as wood shavings, since it does not absorb as well. There should be openings for fresh air, and sufficient space to lie down comfortably. Requirements for swine transportation are described in manuals for transportation of live animals by road[50] and air.[51]

Figure 7 Vari Kennel® transport kennels are well suited for transporting minipigs under 25 kg. (Photo courtesy of Ellegaard Göttingen Minipigs, Soroe Landevej, Dalmose.)

record keeping

Record keeping requires individual identification of animals. Swine may be identified with ear notches, ear tattoos, ear tags, neck bands, or intramuscular electronic transponders. For temporary marking of swine, colored markers or spray can be applied on the skin.

Individual animal records should contain the animal's identification number, identification numbers of the parents (if known), and pen number. Body-weight development and medical treatment should be recorded as well, and preferably also information about feed intake, abnormalities, and research procedures.

securing welfare

During all phases of husbandry, a good level of welfare needs to be secured, meaning that the animals need to be able to cope with their environment at any time. A good level of welfare can be obtained by covering five basic needs, including the freedom from hunger, thirst and malnutrition, appropriate comfort and shelter, the prevention of disease, the freedom from fear, and the freedom to display normal patterns of behavior. For swine, particular attention should be paid to flooring and bedding material, since swine have prominent rooting, grazing, and browsing behaviors. The absence of a suitable substrate for rooting may lead to a compromised welfare, resulting in stereotype behavior such as empty chewing or bar biting. Social grouping is also important for swine. Swine are social animals that live in stable maternal family groups, and should therefore preferably be housed in small groups. Adult males, which can be solitary under natural conditions, can be housed individually. Housing, social grouping, nutrition and environmental conditions should ensure that the animals' capability to cope with their environment is optimized.[52]

management and quality assurance

In order to achieve the highest scientific outcome with the smallest possible implications for the test animals, it is essential to standardize production and assure quality. Quality assurance of laboratory animals is especially concerned with microbiological and genetic factors. Furthermore, the status of swine as an important meat production animal means that certain legal demands must be fulfilled.

microbiological monitoring

Infections in Swine and Their Impact on Experiments

In Chapter 4, information about important infections in laboratory swine are given. Such infections may in various ways have an impact on experiments with swine. Basically, the impact may be divided into the following.

Pathological Changes, Clinical Disease, and Mortality

Clinically apparent disease is most often due to bacterial infections, such as *Actinobacillus pleuropneumoniae*, the cause of a severe lung disease. Animals suffering from such disease are seldom used for experiments, but the infections may of course have severe economic impact on laboratory swine breeding and maintenance. However, some infections induce severe pathological changes in the animals, which may not be discovered until during the experiment. *Mycoplasma* spp. and *Haemophilus parasuis* both cause polyserositis,[53] and surviving swine have adhesions in the thorax and abdomen, which are often not discovered until they have already been anaesthetized and laparat-omized for surgery.

Immunomodulation

The immune system may be modulated by spontaneous infections, also in the absence of clinical disease, an effect which may be either suppressive or activating—or both at the same time, but on different parts of the immune system. As a general rule of thumb all viruses should be regarded as immunosuppressive, one of the reasons being the viremic phase in the pathogenesis of many virus infections during which cells of the immune system may be infected. The immunological impact of the African swine fever virus has been intensively studied, and it has been shown that changes in the immune system persist after recovery from infection.[54] African swine fever is not very common in herds supplying swine for experiments, but more frequently occur-ring viruses, such as influenza viruses, may have serious impact on the immunology of the animals.[55]

Physiological Modulation

Some microorganisms have a specific effect on enzymatic, hema-tological, and other parameters that are monitored in the animal during an experiment. Organic function disturbances may change the outcome of the experiment without the knowledge of the scientist; for example, an altered function of the liver caused by spontaneous hepatic infections may influence pharmacokinetics. In one study it was shown that the most important change in drug elimination dur-ing an acute phase response induced by *A. pleuropneumoniae* is a suppression of oxidative hepatic biotransformation.[56] Also, antibiotics or anthelmintics used to treat infections or infestations may inter-fere with pharmacokinetics. Strong cytochrome oxidase 1A induction, caused by fenbendazole therapy in swine, may negatively affect the efficacy and pharmacokinetics of fenbendazole itself or other simul-taneously or consecutively administered drugs.[57]

Competition between Microorganisms within the Animal

Spontaneous infections in an experimental infection model may com-pete with the experimental infection, which in the worst cases may be impossible to induce. For instance, gnotobiotic or microbiologi-cally defined (see below) swine may be experimentally infected with *Helicobacter pylori*, while this is not possible in conventional swine.[58] Such differences between different microbiological categories of ani-mals may be explained by natural infection with related agents, as species-specific *Helicobacter* spp. have been isolated from swine.[59] Infections with species other than *Helicobacter* may play a role as well.

Interference with Reproduction

For the breeder, changes in fertility of the animals may represent serious trouble. Infections giving rise to clinical disease in a major part of the population are very likely to reduce fertility, and it is also quite obvious that infections leading to high mortality in neonatality, such as rotaviruses,[60] will reduce the outcome of breeding or disturb experiments using newborn pups. Direct effects of the infection on reproduction, such as a change in sex hormones, pathological anatomical changes in the reproductive tract, or infection of the embryo causing abortion and stillbirths, are also observed. This is seen after infection with parvovirus,[61] circovirus,[62] and, in Australia, the Menangle paramyxovirus.[63] Reproduction studies in which the absence of such infections have not been documented may be prone to criticism.

Interference with Oncogenesis

Infectious agents may induce cancer, enhance the carcinogenic effect of certain carcinogens, or reduce the incidence of cancer in laboratory animals; and opportunistic pathogens may become pathogenic in animals with cancer: swine with lymphosarcomas, for example, are more likely to die from colibacillosis than swine without lymphosarcomas.[64]

Contamination of Biological Products

Microorganisms present in the animal may contaminate samples and tissue specimens such as cells, sera, etc. This may complicate in vitro maintenance of cell lines, and may interfere with experiments performed on cell cultures or isolated organs. Further, the reintroduction of such products into animal laboratories will impose a risk to the animals kept in that laboratory. In fact, many of those mycoplasmae that complicate maintenance of cell lines[65] are porcine mycoplasmae originating from either the use of porcine serum in the media or from the porcine origin of the cells to be grown.[66]

Impact of the Normal Flora

If not reared under gnotobiotic conditions, swine, as do other animals, harbor a huge number of gut bacteria. There is scientific evidence that the nature of the gut microbiota—especially in early life—has an impact on the maturation of the immune system and thereby on the development of inflammatory diseases, such as inflammatory bowel disease (IBD) and diabetes, for which swine are used as models.

It is today possible by various molecular techniques to profile the gut microbiota, and such techniques will probably become increasingly applied to document uniform animals from laboratory animal vendors to secure standardization.[67]

Zoonotic Diseases

Some important zoonoses are listed in **Table 10**. It should be noted that zoonoses of porcine origin in laboratory staff are rare. One major risk seems to be the development of erysipelas after contact with swine infected with *Erysipelothrix rhusiopathiae*.[68]

The potential future use of swine as organ donors for human xenotransplantation has created a new term, *xenozoonosis*,[69] which refers to infections that may pass with the organ from the animal donor to the human recipient. Ordinary zoonotic infections known to be present in swine, such as encephalomyocarditis virus[70] and rotaviruses,[71] are probably easy to prevent, so concern has been related to the risk of activating porcine endogenous retroviruses (PERV) in the recipients, which may then spread the infection to relatives and cause a world-wide retroviral epidemic.[72] However, studies on immunosuppressed, xenotransplanted baboons[73] and on patients exposed to living swine tissues have been unable to show such cross-species transmission.[74,75]

Legal Regulations

Most Western countries have regulations concerning import and export of live animals. OIE, originally the International Office of Epizootics, now renamed the World Organisation for Animal Health (www.oie.int), is an organization of countries that register infectious diseases worldwide, set standards, and issue guidelines on how to prevent the spread of infectious animal diseases. OIE is responsible for the administration of the Terrestrial Animal Health Code (TAHC), according to which the member countries have certain responsibilities. One of these responsibilities is the maintenance of an alert system for diseases listed in the code (**Table 11**). Swine are subjected to strict regulations, as pork production is essential for the national economy of several countries. According to U.S. food and drug regulations, the Secretary of Agriculture has authority to make such regulations and take such measures as he or she may deem proper to "prevent the introduction or dissemination of the contagion

TABLE 10 PORCINE INFECTIONS WITH A ZOONOTIC POTENTIAL

| Agent | Symptoms | | Transmission to Humans | Risk for Research Staff |
	In Pigs	*In Humans*		
Brucella suis	Mostly no symptoms. Classical signs are abortions, infertility, weak piglets at birth, infection of the testes, and arthritis	Severe fever and disabling lesions of the spine, which may become lethal	Direct contact with tissues, blood, urine, vaginal discharges, or aborted fetuses and placentas from infected animals	Low
Campylobacter jejuni	Diarrhea or no symptoms	Diarrhea	Fecal contamination of human food	Low
Escherichia coli (only certain strains causing porcine edema disease)	Neurologic symptoms, edema of the eyelids, occasionally edema of the neck and abdominal skin	Similar to the pig, occasionally damage to the heart valves, or sudden death	Fecal contamination of human food	Low
Erysipelothrix rhusiopathiae	Fever, arthritis, often squarish erysipeloid lesions (pinpoints to > 3 inches)	Similar to the pig, occasionally damage to the heart valves, or sudden death	Contact with infected pigs	High
Leptospira spp.	Normally none, in some cases abortions	Hemorrhagic fever	Contact with porcine urine	Low
Salmonella spp.	Diarrhea	Diarrhea	Fecal contamination of human food	High
Taenia solium	None	Brain seizures, muscle pain and weakness, heart failure, sometimes sudden death due to development of cysticercus	Intake of insufficiently cooked pork	Low
Trichinella spiralis	None	Muscle encapsulations resulting in weakness and muscle pain	Intake of insufficiently cooked pork	Low
Mycobacterium bovis, M. avium and *M. tuberculosis*	None (granulomas may be found during necropsy)	Tuberculosis	Inhalation and ingestion	Low

TABLE 11 REPORTABLE INFECTIOUS DISEASES OF SWINE REGISTERED BY THE
INTERNATIONAL OFFICE OF EPIZOOTICS (OIE)

List A Diseases	List B Diseases
Definition:	*Definition:*
Transmissible diseases that have the potential for very serious and rapid spread, irrespective of national borders, that are of serious socio-economic or public health consequence, and that are of major importance in the international trade of animals and animal products	Transmissible diseases that are considered to be of socio-economic and/or public health importance within countries and that are significant in the international trade of animals and animal products
Foot and mouth disease	Anthrax
Vesicular stomatitis	Aujeszky's disease
Rift Valley fever	Leptospirosis
Bluetongue	Rabies
African swine fever	Avian diseases
Classical swine fever	Atrophic rhinitis of swine
	Enterovirus encephalomyelitis
	Porcine brucellosis
	Porcine cysticercosis
	Porcine reproductive and respiratory syndrome
	Transmissible gastroenteritis
	Trichinellosis

of any contagious, infectious, or communicable disease of animals ...
from a foreign country into the United States or from one State ... to
another." Such regulations are a common part of agricultural legisla-
tion in most Western countries.

In order to achieve these goals, national departments of agriculture in the OIE member countries run mandatory health programs
aimed at eradication of OIE reportable diseases, such as brucellosis,
foot and mouth disease, and swine fever. In addition, they allow
the import and export of swine and porcine products only under a
strict state veterinary control. Therefore, prior to import and export
of swine, one should always contact the national or local veterinary
authorities to ensure that licenses are granted and swine are accom-
panied by all necessary certificates.

Precautions to Prevent Infections

In several countries, organizations to ensure the health of farm swine
exist. Since the late 1950s, such national organizations have set up
systems for swine production according to a three-step principle:

1. Production of germ-free animals by the rederivation techniques of caesarian section or embryo transfer.
2. Start-up of a breeding colony on the basis of the rederived animals in a protected (barrier) environment.
3. Regular screening of the colony for the presence of certain agents (health monitoring).

Rederivation

Germ-free swine may be produced by removing them aseptically from the uterus by hysterectomy and rearing them in isolators.[76] In the isolators they may be associated with a specific flora. The term *gnotobiote* refers to germ-free animals as well as animals with a specific flora. Gnotobiotic swine are used for a variety of purposes, including immunological studies of natural and adaptive immunity and early antibody synthesis, monoassociation with certain bacteria, and studies of parenteral nutrition.[77–81] Germ-free swine have a lower count of peripheral blood leucocytes (although an increased number of peripheral T-cells), a lower percentage of neutrophilic granulocytes, a lower total serum protein, and higher serum albumin and beta-globulin levels.[82] An alternative to cesarean section is embryo transfer.[83] The piglets receive the microbiological status of the recipient mother.

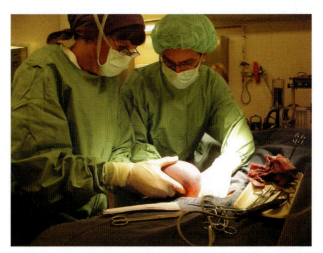

Figure 8 Cesarean section for rederivation of a pig herd. The piglet is released by an aseptic caesarean section and transferred to an isolator, where it is artificially fed and kept warm. (Photo courtesy of Ellegaard Göttingen Minipigs, Soroe Landevej, Dalmose.)

Barrier Protection

As long as rederived swine are kept in isolators, gnotobiotic status can be maintained. For swine used to start up a breeding colony, housed under conditions protecting them against certain agents, and for which a monitoring system is operated, terms such as *minimal disease* (MD), *specific pathogen free* (SPF), and *microbiologically defined swine* are used.

Meat-producing SPF systems consist of various levels of herds, that is, animals derived from caesarean sections are used for starting a primary SPF herd that may in turn deliver swine for breeding in a secondary SPF herd. No further animals are introduced into primary herds.

Swine produced under barrier protection equivalent to what is applied when breeding laboratory rats and mice are called *microbiologically defined*.[84] Microbiologically defined swine have improved the possibilities of using swine for research, not only in toxicology, but also in more specific research, such as studies of *Helicobacter pylori* infection.[85] Microbiologically defined swine seem to differ in serum and hematology values from conventional swine in the same way as described for gnotobiotic swine above.[21]

Some quarantine and barrier demands are also fulfilled in the production of SPF and MD swine, although in general they are not as strict as those for microbiologically defined swine[84] (**Table 12**). The SPF production method was first applied in agricultural swine farming, where it was used for eliminating infections that had a negative

TABLE 12 TYPICAL HOUSING CONDITIONS FOR DIFFERENT CATEGORIES OF PIGS FOR EXPERIMENTAL PURPOSES

	Conventional	Specific Pathogen Free (SPF), Minimal Disease (MD)	Microbiologically Defined
Caesarian originated	–	+	+
Barrier:			
Quarantine regulation	–	+	+
Change of dress	–	+	+
Shower	–	–	+
Decontamination of:			
Diet	–	–	+
Equipment	–	–	+
Water	–	–	+
Absolute filtered ventilation	–	–	+
Health monitored	–	+	+

impact on the farmers' economy. However, because they are free of mycoplasmas, SPF swine have also been widely used for experimental purposes. Within some agricultural organizations, the term *minimal disease* (MD) refers to swine colonies in which some but not all of the agents covered by the regulations of that organization are found. This term may be used for former SPF colonies infected with *Mycoplasma hyopneumonia*. Swine that are neither SPF nor microbiologically defined are called conventional.

Health Monitoring

Table 13 gives some examples of infections covered by the SPF term, although it should be remembered that different organizations in different countries work with slightly different lists. Some associations of laboratory animal science, such as the Federation of European Laboratory Animal Science Associations (FELASA),[86] have issued guidelines for advanced health-monitoring in swine for experiments. Such guidelines recommend monitoring for a range of agents, but as porcine health is a rapidly developing field due to the enormous impact it has on the economy in some countries, these guidelines have to be currently supplemented for new agents. Commercial vendors specialized in purpose-breeding of swine for research have set up their own systems, which may differ slightly from the guidelines. Also, as can be seen in **Table 14**, the approaches for securing a high health standard differ somewhat, as some apply vaccination and antiparasitic treatment, while others do not. Although it might

TABLE 13 INFECTIONS THAT MAY BE COVERED UNDER THE TERM *SPECIFIC PATHOGEN FREE*

Morbus Aujeszky
  Herpesvirus (Aujeszky's virus)
Pneumonia
  Mycoplasma hyopneumoniae
  Actinobacillus pleuropneumonia
Swine dysentery
  Serpulina hyodysenteriae
  Transmissible gastroenteritis
  Coronavirus (TGE)
Atrophic rhinitis
  Pasteurella multocida
Sarcoptic mange
  Sarcoptes scabiei
Porcine reproductive and respiratory syndrome
  PRRS virus

TABLE 14 INFECTIONS INCLUDED IN SOME PROGRAMS FOR HEALTH MONITORING OF MICROBIOLOGICALLY DEFINED PIGS

	FELASA[86]	Ellegaard Göttingen Minipigs[a]	Sinclair Bio Resources[b]
Viral Infections			
African swine fever	If present in country	Cons. abs.	Cons. abs.
Morbus Aujeszky	If present in country	Yes	Yes
Encephalomyocarditis virus	If present in country	Yes	No
Enteroviral encephalomyelitis (Teschen)	If present in country	Cons. abs.	Cons. abs.
Hemagglutinating encephalomyelitis	If present in country	Yes	No
Transmissible gastroenteritis	If present in country	Yes	Yes
Porcine circovirus	No	Yes	No
Porcine cytomegalovirus	If present in country	No	No
Porcine epidemic diarrhea	If disease associated	Yes	Cons.abs.
Porcine influenza	If present in country	Yes	Yes
Porcine parvovirus	If present in country	Yes	Vaccination
Porcine reproductive & respiratory syndrome	If present in country	Yes	Yes
Porcine rotavirus	If present in country	Yes	No
SMEDI (enterovirus)	If present in country	Cons. abs.	No
Swine fever	If present in country	Yes	Cons. abs.
Vesicular stomatitis	On request	Cons. abs.	Yes
Bacterial and Fungal Infections			
Actinobacillus pleuropneumoniae	Yes	Yes	Vaccination
Actinobaculum (Eubacterium) suis	Yes	Yes	No
Actinomyces pyogenes	On request	No	No
Brachyspira (Serpulina) hyodysenteriae	On request	Yes	No
Bordetella bronchiseptica	Yes	Yes	Vaccination
Brucella suis	On request	Cons. abs.	Yes
Candida albicans	No	Yes	No
Campylobacter spp.	No	Yes	No
Clostridium perfringens	On request	Yes	No
Erysipelothrix rhusiopathiae	Yes	Yes	Vaccination
Helicobacter spp.	No	Yes	No
Haemophilus parasuis	Yes	Yes	Vaccination
Klebsiella pneumoniae	No	No	No
Lawsonia intracellularis	No	Yes	No
Leptospira spp.	Yes	Yes	Yes/ Vaccination
Listeria monocytogenes	No	Yes	No

TABLE 14 (CONTINUED) INFECTIONS INCLUDED IN SOME PROGRAMS FOR
HEALTH MONITORING OF MICROBIOLOGICALLY DEFINED PIGS

	FELASA[86]	Ellegaard Göttingen Minipigs[a]	Sinclair Bio Resources[b]
Mannheimia (Pasteuralla) haemolytica	No	Yes	No
Microsporum spp.	On request	Yes	No
Mycoplasma hyopneumoniae	Yes	Yes	No
Pasteurella multocida	Toxin prod.	Yes	Vaccination
Salmonellae	Yes	Yes	Yes
Staphylococcus hyicus	Yes	Yes	No
β-hemolytic streptococci	Yes	Yes	Yes
Streptococci (beta-hemolytic)	No	Yes	No
Streptococcus pneumoniae	No	Yes	No
Streptococcus suis	Yes	Yes	No
Trichophyton spp.	On request	Yes	No
Yersinia enterocolitica	Yes	Yes	No
Parasitological Infections			
Arthropods	Only sarcoptes	Yes	Yes*
Helminths	Yes	Yes	Yes*
Coccidiae	Yes	Yes	Yes*
Toxoplasma gondii	Yes	Yes	No

On request: The guidelines demand that the producer screen for the agent if requested by the user.
Cons. abs.: Considered absent; some agents are not included in the program, being irrelevant because of national health programs.
If present in country: The infection should be included in the health-monitoring program only if it has not been eliminated by national health programs.
[a] Göttingen minipigs; http://www.minipigs.com/
[b] Sinclair, Yucatan and Hanford minipigs; http://www.sinclairbioresources.com/
* Antiparasitic treatment is applied

make a breeder's health report more impressive, it might make less sense to include OIE organisms already controlled by national regulations. Advantages and disadvantages of these different approaches depend on individual judgment based upon purpose of research, type of experimental facilities, etc. Most standards for health monitoring also demand regular clinical inspections by a trained veterinarian. Prior to any purchase of swine for experimental purposes, a health report should be obtained from the vendor.

In breeding colonies health monitoring is based on random representative samples. It is assumed that though only a few animals can be sampled for examination, the results can be used to describe

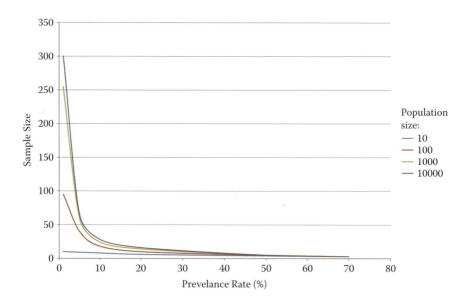

Figure 9 Sample size for health monitoring in relation to the expected prevalence of the monitored infection. If the colony size is small, i.e., fewer than 10,000, or the expected prevalence low, i.e., under 20%, the population size has some impact on the sample size. If this is not the case, the sample size (S) can be calculated from the following equation: $S \geq \log C/(1 - (p*N1))$. C = accepted risk of a false negative diagnosis; N1 = sensitivity of the applied test; p = expected prevalence.

the entire colony. If one animal is found to be infected with a certain organism, the entire colony is considered infected. Often the opposite is also assumed, that is, if no animals are found to be infected with a certain organism, the entire colony is free of that particular organism. This, however, requires that a statistically valid sample size has been used.[87] Such a sample size may be calculated according to **Figure 9**.

As soon as a sample has been taken, it becomes historical. In a comparison of two samplings taken at a certain interval, the last sampling will show whether changes have occurred between the two samplings. A random sample of animals may be predictive for the microbiological status of other animals only if these animals have some kind of contact with one another. Therefore, it is essential to define the microbiological entity, that is, to have a definition of the group of animals for which a sample is predictive. Each simple one-room unit used for breeding at a commercial vendor will normally have its own staff and barrier to separate it from all other units, and therefore each of these separate units should be regarded as separate microbiological entities. However, if a unit is separated into different rooms that are all served by the same staff and get their supplies through a common barrier, it is difficult to define the microbiological entity. But it should be noted that the more subunits included, the higher the uncertainty of the sampling, and if the infection is found only in one subunit, the prevalence is diluted, which should be taken into consideration when calculating the sample size.[88]

Common Findings in Health Monitoring

In the majority of conventional swine, antibodies against parvo- and influenza viruses as well as *H. parasuis* are found. Less commonly, antibodies to porcine coronaviruses, hemagglutinating encephalomyelitis virus, porcine reproductive and respiratory disease viruses, and all serotypes except for type 7 and 8 of *A. pleuropneumoniae* and *Mycoplasma hyopneumoniae* are found. Bacteriological, fungal, and parasitological examinations often reveal the presence of *A. pleuropneumoniae*, *Bordetella bronchiseptica*, *Clostridium perfringens*, *Eubacterium suis*, *H. parasuis*, *Pasteurella multocida*, *Salmonellae*, *Staphylococcus hyicus*, and *S. suis*, as well as helminth eggs and coccidiae.[89] SPF swine do not harbor pathogens such as *A. pleuropneumoniae* or *Mycoplasma hyopneumoniae*, which are mentioned in the guidelines for this category of swine, but as for non-listed agents, there is no guarantee for the absence of the other above-mentioned agents. Swine housed according to the principles for microbiologically defined animals have been shown to harbor a lower number of infections than swine housed according to agricultural SPF guidelines; but even in microbiologically defined minipigs, health monitoring may reveal the presence of certain agents with a research-interference potential, such as rotaviruses, *Campylobacter* spp., and β-hemolytic streptococci.[90]

genetic monitoring

Genetic monitoring as a tool for standardization in laboratory rodent breeding has until recently not been applied in the same way in the production of swine for research. Most genetic programs are aimed at the development of research swine more suitable than at present. For this purpose, some breeders register such parameters as[84,91]

1. Weight at birth
2. Growth rate

3. Number of visible ear veins

4. Temper

5. Number of teats

6. Malformations

Other parameters may be used as well, such as small body size in an attempt to breed even smaller minipigs.[28,29]

The development of transgenic and cloned swine has obviously introduced the same need for documenting transgenes and targeted mutations in swine as is done in mice. This is typically done by polymerase chain reaction (PCR), or in some cases by southern blot. Gene maps of the entire genome of swine have been established.[92]

accreditation

Certain bodies may accredit laboratory animal facilities with respect to either animal welfare or production quality. The successful introduction of a quality system confers peer-recognition against an independent standard, thereby providing assurance of standards of animal care and use and improving the quality of animal studies.[93] For registering or licensing non-clinical health and environmental studies, good laboratory practice (GLP) is a legal requirement. GLP guidelines are often relevant only for institutions performing those types of studies.

AAALAC International

The Association for Assessment and Accreditation of Laboratory Animal Care (AAALAC) runs a voluntary accreditation program in which research institutions and animal breeders demonstrate that they are not only meeting the minimums required by law, but are going the extra step to achieve and showcase excellence in animal care and use. Along with accreditation, AAALAC also offers independent assessments of animal research programs to help institutions continue to improve their animal care and use practices. AAALAC is an international organization, with offices in the United States, Europe, and Asia.

Breeding of the most common laboratory animal species is subject to regulations on designated breeding of research animals in European Union member states, which means that all breeders must be licensed by and work under the inspection of the local animal welfare authorities. These regulations, however, do not apply to swine breeding; therefore, swine breeders especially may benefit from AAALAC accreditation.

In the United States, Institutional Animal Care and Use Committees (IACUC) are charged with oversight and evaluation of animal care and use under the terms of the Animal Welfare Act and the *Guide for the Care and Use of Laboratory Animals.* The quality and effectiveness of the IACUC's performance are routinely assessed during AAALAC site visits.[94]

ISO 9000

ISO 9000 is a set of universal standards for a Quality Assurance system issued by the International Organization for Standardization (ISO), a worldwide federation of national standards bodies. The system should assure connection between the quality of the product and the expectations of the customer. ISO-9000-registered vendors have ensured that they have a proper quality assurance system. This system is becoming increasingly important, especially in Europe. The overall accreditation is granted by the national ISO member organization. In the United States, this is the American National Standards Institute.

veterinary care

The veterinary care of swine starts with the arrival of the animals at the facility. Clinical examination should be a standard procedure upon arrival. Swine purchased from agricultural sources usually have no health record, and microbiological sampling may be included as part of the clinical examination. Swine originating from specialized laboratory swine breeders are usually provided with a health monitoring report (HMR). The HMR should be studied for positive findings, and it should be considered whether positive findings pose a risk for the existing swine population in the facility. Until results from the clinical examination are available, swine should be housed in quarantine facilities.

clinical examination of swine[95,96]

A good clinical examination starts with the history of the individual animal. Breed, sex, age, weight, and the animal's identification number should be recorded, and existing HMR or animal certificates should be added to the record.

Swine are usually not accustomed to handling, so to avoid misleading observations of physiological parameters, swine should be observed as much as possible without handling or restraint. However, swine arriving at the facility are heavily influenced by transportation, and it should be taken into account that the animals are excited by the new environment. Posture, behavior, and respiration should be observed prior to restraint.

Diseased swine may have a long, rough-haired coat, with the hair standing up (piloerection). A dog-sitting posture is associated with respiratory difficulty. Hypothermic swine usually lie in sterna recumbence, with the legs tucked under the body. Normally, swine lie on their side, but this is avoided by animals with heart disease or abdominal pain. Lame swine are reluctant to rise. Swine in a good nutritional state should neither be too fat nor have observable bony structures in the body. The normal body temperature is 38–39°C (100–102°F).

A checklist for clinical examination of swine is presented in **Table 15**.

microbiological sampling[71,95]

Clinical examination may include microbiological sampling, especially of swine without a known health status. However, when swine are purchased from a reliable source, microbiological sampling is often not performed.

Blood samples for serology are taken from the jugular vein or cranial caval vein. From ear veins, usually not more than a few drops of blood can be taken, but this may be sufficient for some tests. Methods of blood sampling are described in Chapter 5.

Sterile cotton-tip swabs should be used for sampling nasal secretions. The swab is introduced into one of the nostrils and inoculated onto a plate or kept in Stuart's transport medium for bacterial culture.

A fecal sample is collected from the rectum and divided into two parts. One part is kept in a sterile tube for microbiological examination, while the other part, used for parasitology, is kept in a plastic container for fecal flotation.

A skin scraping for fungi and mites is taken with a sterile scalpel blade from an area with lesions or the shoulder and kept in a sterile tube for microscopic inspection.

The samples are immediately sent to a health-monitoring laboratory for serological, microbiological, and parasitological examination. Infectious agents that can be included in the examination are listed in **Table 14** in Chapter 3

diseases of swine

A complete overview of diseases of swine is not possible within the scope of this book. If a detailed description of disease is required, it is recommended to use a reference handbook.[95,96] *An Atlas of Swine Diseases*[97] is a useful tool in diagnosing disease. A summary of swine diseases is presented in **Table 16**.

TABLE 15 CHECKLIST OF THE CLINICAL EXAMINATION OF SWINE

Condition	Desirable	Undesirable
body weight	normal for breed, age, and sex	thin, fat, or obese
skin	flat hair coat with a normal thickness for the breed, white-pink color in unpigmented areas of the skin	pale, anemic skin, thickened keratinized patches, wounds and sores, any lesion
mouth, nose, and eyes	no discharge, pink mucosa, normal dentition for age	discharge, blindness, lesions, ulcers
anus, vulva	normal size, pink mucosa, not fouled; vulva may show watery, clear to whitish discharge	diarrhea, excessive or purulent discharge from vulva, prolapses; red, swollen vulva is a sign of estrus
feet, legs, and gait	normal stance and movement	foot lesions, swollen joints, difficulty in lying down and getting up, lameness
respiration	no coughing or sneezing, normal respiration	frequent coughing or sneezing, forced or superficial respiration
cardiovascular system	normal pulse rate	collapse during restraint
feces	solid, brownish-yellow to brownish green color, depending on feed	watery or extremely hard, reddish, black or yellow color

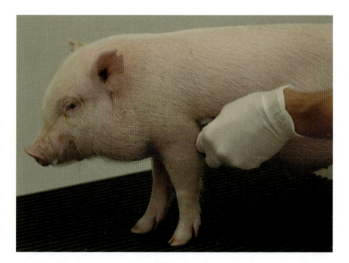

Figure 10 Auscultation can be difficult in swine because of coarse hair, and often because of lack of cooperation and loud vocalization of the animal. In young minipigs, though, this it is often not the case. (Photo courtesy of Ellegaard Göttingen Minipigs, Soroe Landevej, Dalmose.)

TABLE 16 DISEASES OF SWINE[95,96]

System	Clinical Sign	Cause	Remarks
Cardiovascular	Myocarditis	Streptococci	Increased heart rate, sudden death
		Encephalomyocarditis virus	
		Foot-and-mouth	
		Swine vesicular disease	
		Vitamin E/selenium deficiency, iron deficiency	
		Malignant hyperthermia	
	Pericarditis	*Pasteurella* spp.	Increased heart rate, sudden death
		Mycoplasma spp.	
		Hemophilus parasuis	
		Actinobacillus pleuropneumoniae	
		Streptococci	
	Vasculitis	African swine fever	Blood loss, bleedings, anemia
		Hog cholera	
		Erysipelas	
		Hemophilus parasuis	
		Actinobacillus pleuropneumoniae	
	Hemorrhagic conditions	Intoxication	Blood loss, bleedings, anemia
		Factor VIII deficiency	
		African swine fever	
		Platelet disorders (thrombocytopenia)	
	Anemia	Iron deficiency	Pale skin, lethargy
		Gastric ulcers	
		Parasites	
Digestive	Enteritis	Rotavirus	Watery to pasty diarrhea
		Coronavirus	Watery diarrhea
		Isospora suis (coccidiosis)	
		Escherichia coli	
		Clostridium perfringens	Hemorrhagic diarrhea
		Serpulina hyodysenteriae	Mucohemorrhagic diarrhea
		Salmonella spp.	
		Trichuris suis	
		Campylobacter spp.	Variable

TABLE 16 (CONTINUED) DISEASES OF SWINE[95,96]

System	Clinical Sign	Cause	Remarks
Respiratory	Atrophic rhinitis	*Pasteurella multocida*	Nasal discharge, snout deformities
		Bordetella bronchiseptica	Nasal discharge, sneeze
		Mycoplasmae	Nasal discharge
	Pneumonia	*Mycoplasma hyopneumoniae*	Dry cough
		Actinobacillus pleuropneumoniae	Forced respiration, fever, dog-sitting position
	Pleuritis	*Bordetella bronchiseptica*	Cough, forced respiration, fever
	Viral airway infections	Influenza H1N1 and H3N2	Dry cough, superficial respiration
		Aujeszki's disease	Cough, periodical fever
Skin and skeletal	Lameness	Osteochondrosis (genetic, vitamin D deficiency, nutrient imbalances)	Weak legs, kneeling position
	Arthritis	Streptococci, mycoplasmae	Swollen, hot joints
	Foot lesions	Housing conditions (concrete floors, grids)	Lesions on hooves, lameness
	Exudative dermatitis	*Staphylococcus hyicus*	Greasy pig disease, black greasy spots
	Erysipelas	*Erysipelothrix rhusiopathiae*	Red spots, generalized
	Abscesses	Streptococci, *Actinomyces pyogenes*, multiple agents	Enlargement under skin, purulent inflammation
	Dermatitis	*Candida albicans*	Circular spots, gray exudate
	Ringworm	*Microsporum, Trichphyton*	Circular red spots
	Swine pox	Swine poxvirus	Palules and pustules
	Vesicular diseases	Foot-and-mouth, swine vesicular disease, parvovirus	Vesicular lesions especially on the snout and edge of the hooves
	Parakeratosis	Zinc deficiency	Erythematous areas, overlaid with scales
	Hyperkeratinization	Unknown	Brownish greasy scales on the back

It should be stressed that prevention of disease by procuring swine from reliable sources is essential. Use of health-monitored swine, preferably with an SPF or microbiologically defined status, reduces the risk of introducing infectious disease into the laboratory facility. Swine from sources with differing health statuses should not be housed together. Thorough sanitation and disinfection of animal facilities is essential for disease control. Pathogens may be harbored in the animal facility; consequently, non-diseased swine may become infected even though no swine have been present in the facility for a long period.

therapeutic treatment

Medical treatment starts early in the life of swine. All newborn swine are treated with a prophylactic iron medication. Leaving out this iron treatment results in severe anemia,[98] because of the rapid growth rate of newborn swine and the low iron content of sow's milk. Most common is an intramuscular treatment with 100 mg/kg iron-dextran,[99] but oral dosing is also applied to prevent neonatal anemia.

One should consider that the use of therapeutics may have an influence on the experiment in which the animals are being used, and iron treatment will often lead to siderosis, with iron deposits in liver, kidney, and lymph nodes.[99,100] Swine, especially ones originating from agricultural sources, are also routinely treated prophylactically with antimicrobial and anthelminthic agents, and such treatments could lead to disease after experimental procedures, since infection, suppressed with these agents, may progress to disease if the animal is immunosuppressed by the experimental procedure.[91]

A relatively small number of pharmaceutical compounds are available specifically for swine. Antimicrobial feed additives make up by far the largest volume of chemotherapeutic agents administered to swine, though these are not typically used in laboratory swine. They are administered at low, often subtherapeutic levels for disease control as well as increased growth rate,[96] and thus they are often called growth promoters. They should not be used in purpose-bred laboratory swine, and are not typically used in miniature swine.

Antimicrobial and anthelminthic therapy may be used in all swine for the treatment of clinical disease. **Table 17** lists antimicrobial drugs for swine, with guidelines for dosing. It is recommended that drug labels be referred to for specific doses.

TABLE **17** ANTIMICROBIAL DRUGS FOR SWINE[96]

Drug	Maximum Daily Dose (mg/kg)[a]	Interval[b]	Route[c]
Amoxicillin	11–13	SID	PO
Ampicillin	20	TID	SC, IM
Erythromycin	11–20	SID	IM
Lincomycin	11	SID	IM
Neomycin	11	SID	PO
Oxytetracycline	7–11	SID	IM
Oxytetracycline	44–55	SID	PO
Sulfachlorpyridazyne	77–110	SID	PO
Sulfamethazine	240	SID	PO
Sulfathiazole	220	SID	PO
Tiamulin	9	SID	PO
Tylosin	18–22	BID	IM
Ceftiofur	3	SID	IM
Enrofloxacin	2.5	SIC	IM

[a] Review drug labels for specific dosages.
[b] SID = once a day; BID = twice a day; TID = three times a day
[c] PO = oral (*per os*); SC = subcutaneous; IM = intramuscular

pathological examination

Diseased swine with no prospects of natural recovery or recovery after medical treatment must be euthanized. Methods of euthanization are described at the end of this chapter. General conditions in which euthanization is advised are

- rectal or vaginal prolapses with tissue damage
- wounds of such severity that healing is in doubt
- multiple sores and wounds, including tail-biting wounds
- cardiac failure
- lameness from a broken leg or arthritis
- paralysis or neurological disorders
- hernias
- emaciation
- severe septicemia, pneumonia, or enteritis

The procedure for necropsy is described in Chapter 5.

anesthesia and analgesia

Swine can be anesthetized for several hours without higher risk of complications compared with other laboratory animals. However, some difficulties and pitfalls exist in anesthetizing swine, and thus careful planning prior to anesthesia is necessary. Premedication with sedatives is recommended to avoid the need for physical restraint, as restraint may induce stress. Furthermore, sedation makes it possible to place an ear catheter, necessary for the induction of intravenous anesthesia. Apnea is common, the porcine lung capacity being low compared to other animal species such as dogs and cats, so tracheal intubation is recommended, particularly for long-lasting anesthesia—though for untrained staff tracheal intubation in swine can be difficult. Swine have poor thermoregulatory ability, so warming blankets should be used to prevent hypothermia. Furthermore, some breeds are prone to malignant hyperthermia when subjected to inhalation anesthesia with halothane. Successful anesthesia of swine includes sufficient analgesia, sedation, and muscle relaxation.

Acclimatization

Swine should never be anesthetized during the first days after transportation, since this may increase the risk of death under anesthesia and the risk of postoperative complications. Swine should be acclimatized for one to two weeks in the experimental unit before any anesthesia is administered or surgical procedure is performed. During the acclimatization period, swine should be kept under quiet conditions to avoid stress. Optimally, swine should stay in their own pen until after sedation.

Clinical Examination Prior to Anesthesia

Swine should be clinically examined prior to anesthesia. This can be done shortly before premedication. The clinical examination should be performed quickly, and only in cases where there are signs of disease is supplementary clinical examination needed. A quick examination includes behavioral observations. The respiration rate should also be noted, as high frequency may indicate lung disease or stress. In case of any signs of disease, the clinical examination should be supplemented with auscultation of the heart and lungs and measurement

TABLE 18 CLINICAL EXAMINATION OF SWINE
PRIOR TO ANESTHESIA

Respiration rate	12–18 respirations per minute
Heart rate	94–116 beats per minute
Body temperature	39°C

of the body temperature (**Table 18**). Acute systemic diseases exclude the use of the swine from any anesthesia procedure.[101–103]

Preoperative Fasting

Swine have to be fasted from 6 to 12 hours prior to anesthesia (neonates only 3 hours), even though vomiting rarely occurs in swine. The intestinal transport time in the upper gastrointestinal tract is rapid, but swine have a pouch called the *torus pyloricus* near the pyloric sphincter, which has the consequence that even after 6 to 12 hours of fasting, the stomach still contains remnant food.[101] For surgery involving the abdominal organs, swine have to fast from 12 to 24 hours. An overloaded ventricle can pressure the diaphragm, decrease lung function, and complicate abdominal organ surgery. Also remember to remove the bedding from the cage, because fasting swine will readily consume it. To prevent dehydration, tap water may be provided up until the time of anesthesia. However, fasting swine are often moderately dehydrated, so saline infusion during anesthesia is recommended.

Premedication

Swine are premedicated in order to sedate and relax them prior to anesthesia.[101–103] Some drug combinations provide good analgesic effects during the first hours of anesthesia. Often atropine is added because of its anticholinergic effect of preventing salivation and tachycardia during intubation and anesthesia. Acepromazine, commonly used as a sedative in swine, acts synergistically with other anesthetics and narcotic analgesics, reducing the required dosages of these drugs. The duration of action may extend into the postoperative phase, making for a smooth recovery. However, moderate hypotension may occur. Azaperone is more potent than acepromazine. It has sedative, but no analgetic effects. Benzodiazepines, such as diazepam and midazolam, induce mild hypnosis, sedation, and muscle relaxation, but with only slight analgesia. Diazepam and midazolam potentiate the action of most

TABLE 19 EXAMPLES OF SINGLE PREMEDICATIONS FOR
SWINE[102–105]

Drug	Dose	Comments
Azepromazine	0.1–0.45 mg/kg IM	Slow onset of action
Atropine	0.05 mg/kg IM	Only anticholinergic effect
Azaperone	5–8 mg/kg IM	Moderate sedation
Diazepam	0.5–2 mg/kg IM	Effective muscle relaxation

TABLE 20 EXAMPLES OF PREMEDICATION COMBINATIONS FOR SWINE[102–105]

Drug Combinations	Dose	Comments
Azaperone	4 mg/kg IM	Moderate to deep sedation,
Midazolam	0.5 mg/kg IM	without analgesia
Atropine	0.04 mg/kg IM	
Ketamine	6–10 mg/kg IM	Deep sedation with analgesia
Medetomidine	60–80 µg/kg IM	Short term surgery possible
Butorphanol	0.2 mg/kg IM	
Tiletamine/Zolazepam	4.4 mg/kg IM	Anesthesia for minor procedures
Xylazine	2.2–4.4 mg/kg IM	
Ketamine	10 mg/kg IM	Useful for scanning studies
Midazolam	1 mg/kg IM	
Zoletil mixture (1 bottle Zoletil + 6.25 mL xylazine (20 mg/mL) + 1.25 mL ketamine (100 mg/mL) + 2.5 mL butorphanol (10 mg/mL))	1 mL/15 kg IM	Deep sedation and anesthesia for minor procedures Analgesia for up to 4 hours

anesthetics and narcotic drugs. Midazolam is more potent than diazepam, but has a shorter duration of action. **Tables 19** and **20** present examples of drugs useful in the premedication of swine.

While swine are still in familiar surroundings, premedication should be injected intramuscularly (IM). The procedure should be performed without creating any fear or stress.[102,103] This can be accomplished by offering the swine a small amount of food, or—for young swine—letting an assistant carry the animal during injection. Though it is also possible to restrain adult swine manually, it is not recommended, since this would cause stress and increase the risk of complications during anesthesia. An IM injection can easily be performed 2–3 cm behind the ear by using a syringe, an infusion line, and an 18- to 21-G needle (**Figures 11** and **12**). The needle has to be at least 30, and preferably more than 40 mm in length. Shorter needles may result in injections into subcutaneous fatty tissue, delaying drug absorption and effects. Injection in subcutaneous fatty tissue is a common problem during premedication of swine.

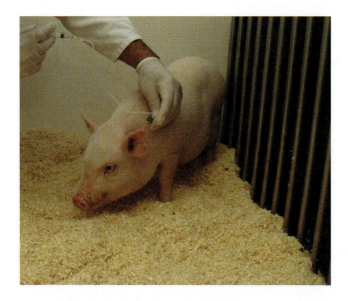

Figure 11 Intramuscular (IM) injections in swine are performed by use of a syringe, an infusion line, and a 30–40 mm, 18- to 21-G butterfly needle. The butterfly needle is inserted just behind the ear. (Photo courtesy of Ellegaard Göttingen Minipigs, Soroe Landevej, Dalmose.)

Figure 12 After the needle is placed, the syringe can be emptied without restraining the animal. Ensure that the volume in the syringe compensates for the dead volume of the infusion line. (Photo courtesy of Ellegaard Göttingen Minipigs, Soroe Landevej, Dalmose.)

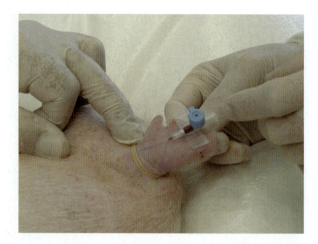

Figure 13 Ear veins are the only accessible superficial vessels in swine. A rubber band is placed tightly around the base of the ear, and an over-the-needle plastic catheter is placed in the ear vein. The catheter is fixed with tape. (Photo courtesy of Ellegaard Göttingen Minipigs, Soroe Landevej, Dalmose.)

Ear Vein Catheter

The ear veins on the dorsolateral surface are the only useful superficial veins in swine.[103] Compared to Landrace swine, Göttingen minipigs and many other minipigs have small ear veins, and this is especially a problem when they have to be cannulated repeatedly. Catheters can be placed in the central or ventrolateral ear vein, while the major artery, located on the dorsolateral area of the ear, must never be used for injection of anesthetics. A rubber band is placed tightly around the base of the ear to enlarge the ear veins. The ear can also be warmed up with warm water. After this, an over-the-needle catheter (e.g., Venflon, 20–22 G) is placed in the ear vein. The rubber band is then removed, and the catheter is flushed with saline. This is done to avoid the formation of blood clots and to ensure that the catheter is correctly placed in the vein prior to medication. The catheter is fixed by adhesive tape (**Figure 13**). If the catheter has to be used for several hours, it can be flushed with a heparin saline solution.

Induction of Anesthesia

Some premedication has such strong sedative effect that further induction of anesthesia is not needed.[102,103] However, in most cases, anesthesia must be induced by intravenous (IV) injection of an anesthetic.

TABLE 21 DRUGS USEFUL FOR INDUCTION OF ANESTHESIA IN SWINE[102-105]

Drug	Dose*	Comments
Propofol	1.5–5.0 mg/kg	Risk of apnea
Thiopenthal	6–25 mg/kg	Strictly given IV
Ketamine	20 mg/kg	Poor muscle relaxation
Midazolam + Ketamine	1.25 + 6 mg/kg	Useful for scanning studies

* The doses should be given until effect (absence of corneal reflexes and good muscular relaxation).

Short-acting anesthetics are preferred for induction of anesthesia. Propofol and thiopentone are the drugs of choice for induction of anesthesia in swine, although a mixture of midazolam and ketamine can also be used (**Table 21**). Propofol is a fast-acting agent with a short recovery time. Thiopentone has properties similar to propofol, but must be given strictly IV to avoid necrosis of the ear tissue. Anesthetics for induction are given until effect is observed, which includes the absence of corneal reflexes and good muscular relaxation (making intubation possible). Instead of injection anesthetics, laryngeal masks can be used for induction of anesthesia with inhalation anesthetics, thus eliminating the requirement for IV catheters.

Endotracheal Intubation

The importance of endotracheal intubation in swine cannot be overemphasized. For long-lasting anesthesia (more than one hour), tracheal intubation is vital, and even for short-term anesthesia it is strongly recommended. Endotracheal intubation maintains a clear airway and an artificial control of respiration. It also protects the airways from aspiration of foreign material. Because of the anatomy of the porcine head, endotracheal intubation is difficult compared to the same procedure in most other large animal species. Therefore, proper training and equipment are needed. A laryngoscope with a long blade (200–250 mm, minimum 195 mm), tubes (**Table 22**), local anesthetic spray (Xylocaine, 2% lidocaine), a syringe with air, and tape for fixation of the tube are needed.[102-105] Swine can be intubated in dorsal, lateral, or sterna recumbence. The method that best serves

TABLE 22 TUBE SIZE FOR SWINE[105]

Body Weight	Tube Size
Piglets	3–5 mm
10–15 kg	5–7 mm
20–50 kg	8–10 mm
Large swine	10–18 mm

Figure 14 Endotracheal intubation in sternal recumbence. The jaws are held open with cotton ropes by an assistant. The tip of the laryngoscope is passed into the pharyngeal cavity and used to displace the epiglottis from the soft palate. (Photo courtesy of Ellegaard Göttingen Minipigs, Soroe Landevej, Dalmose.)

the staff should be chosen. Sterna recumbency appears to be best for persons who have no experience, and therefore intubation with the animal in sterna recumbence will be discussed here. The jaw is held open by an assistant with a cotton rope or gauze strip. Alternatively a mouth gag can be used. The head should be only slightly extended: excessive extension can occlude the airways and make the laryngeal opening more difficult to identify. The tip of the laryngoscope is passed into the pharyngeal cavity and used to displace the epiglottis from the soft palate (**Figure 14**). The laryngeal opening is sprayed with Xylocaine (approximately 1 mL) to prevent spasms. Insufficient anesthesia, excessive manipulation, and repeated unsuccessful attempts to intubate may lead to laryngospasms. The endotracheal tube, stiffened by a guide, is used to press the epiglottis forward onto the base of the tongue, and the tip of the laryngoscope blade is placed in such a way as to bring the vocal cords into view (**Figure 15**). The tube is then advanced into the trachea during expiration. A slight rotation of the tube will facilitate the introduction without undue force. Resistance should not be felt. A free passage of air should be felt when the pig is properly intubated. Any signs of cyanosis or gasping indicate improper tube placement. Only cuffed tubes should be used for large swine, and the cuff must be filled up with air by use of a syringe. Finally, the tube is fixed with tape to the upper jaw. Over-inflation of the endotracheal cuff can cause swelling and edema of the airways, and this may lead to airway obstruction.

Figure 15 View into the pharyngeal cavity. The epiglottis is pressed down by the blade of the laryngoscope, and the arytenoid cartilage is just visible over the edge of the laryngoscope. The tube has to be inserted rostrally to the arytenoid cartilage. (Photo courtesy of Ellegaard Göttingen Minipigs, Soroe Landevej, Dalmose.)

Cricothyrotomy

As an alternative to intubation, cricothyrotomy can be performed very quickly, but is not recommended for animals that have to survive. Two fast methods have been evaluated in minipigs.[107]

Artificial Ventilation

Swine can respire spontaneously during anesthesia. However, for prolonged anesthesia or when the risk of apnea is high, artificial ventilation is recommended. **Table 23** presents ventilation settings for swine. As an example, a 30-kg pig can be ventilated with a tidal volume of 300 mL at 15 times per minute with 100 mL/min oxygen and 200 mL/min air or nitrous oxide.

TABLE 23 VENTILATION SETTINGS FOR SWINE

Parameters	Setting
Tidal volume	8–12 mL/kg
Ventilation frequencies	10–20 respirations per minute
Maximum pressure	20–25 cm H_2O
O_2 : medical air	1 : 2.2
O_2 : N_2O	1 : 2

Thermal Support

Swine are susceptible to hypothermia due to the relatively hairless skin and administration of anesthetics that induce peripheral vasodilation. The body temperature should always be monitored in anaesthetized swine, and blankets or heating pads should be used. IV infusion of warmed saline and an increased room temperature can also prevent hypothermia.

Maintenance of Anesthesia

Anesthesia can be maintained with inhalation or injection anesthesia.[108] Traditionally, inhalation anesthesia has been preferred for long-lasting anesthesia or for studies where swine have to recover from anesthesia. However, expensive equipment is needed for inhalation anesthesia, and, for reasons of occupational health, the equipment has to be controlled periodically.

Inhalation Anesthesia

In general, inhalants are very safe for anesthesia of swine.[109,110] Inhalation anesthesia is recommended for long-lasting anesthesia. Both halothane and isoflurane have been used extensively in swine. The MAC (mean alveolar concentration) values in **Table 24** provide a guideline for the concentration required for anesthesia. However, individual variations exist from pig to pig, and thus anesthetics must always be dosed after effect, that is, until the absence of reflexes is observed.

Halothane

Halothane has a depressant effect on the myocardium, and it sensitizes the heart to catecholamine-induced arrhythmias. In contrast

TABLE 24 MEAN ALVEOLAR CONCENTRATIONS (MAC) OF INHALATION ANESTHETICS IN SWINE[102,103]

Drug	MAC Value
Halothane	0.9–1.3%
Isoflurane	1.2–2.0%
Desflurane	8.3–10.0%
Sevoflurane	2.0–2.7%
Enflurane	1.7%
N_2O	162–277%

to most other inhalants, halothane is partly metabolized in the body, and it may have a hepatotoxic effect.

Isoflurane

Isoflurane is very useful in swine because of its broad safety margin, and the depth of anesthesia is easily regulated.[109] Compared to halothane, it results in a more rapid induction and recovery because of its lower solubility in blood. Halothane and isoflurane have physical similarities, and so can be administrated from the same vaporizers. Isoflurane offers only minimal analgesic effects and is therefore often combined with a strong analgesic, such as fentanyl.

Desflurane

Desflurane, which is an isoflurane isomer, has a low solubility in blood, and therefore induction and recovery are both very fast. The anesthesia depth can easily be regulated. Desflurane is more volatile than halothane and isoflurane, so special vaporizers are needed, adding expense that often limits the use of desflurane in swine.

Sevoflurane

Sevoflurane, which is also an isomer of isoflurane, is a useful anesthetic for swine. It may be the choice for high-risk cases, such as young and sick swine. It has a fast recovery.

Enflurane

Enflurane is, like desflurane and sevoflurane, an isoflurane isomer. Premedication is needed, since enflurane should be avoided for induction because of a relatively slow uptake. High concentrations of enflurane can cause a seizure-like response in swine.

Nitrous Oxide

Nitrous oxide (N_2O) is not a potent anesthetic, and therefore it must be administrated in high concentrations, typically 66%. To avoid hypoxia, 75% N_2O and 25% O_2 is the maximal N_2O concentration for safe anesthesia. Nitrous oxide is used only as a supplement to other inhaled or injectable anesthetics. It reduces the dose of more potent anesthetics (approximately 50% reduction in isoflurane), and its low toxicity and minimal depression of cardiovascular and respiratory systems increase safety and maintain the swine in a better physiological state. There is a risk of hypoxia after ending the anesthesia due to the rapid movement of N_2O from blood to alveoli. Therefore, swine should breathe pure O_2 during the first 10 minutes after

TABLE 25 COMBINATION OF INHALATION ANESTHETICS IN SWINE

1.0–2.0% Halothane in a nitrous oxide/oxygen (1/1) mixture

1.5–2.5% Isoflurane in a nitrous oxide/oxygen (1/1) mixture

2.5–3.5% Sevoflurane in a nitrous oxide/oxygen (1/1) mixture

discontinuing N_2O anesthesia. Combinations of inhalation anesthetics are presented in **Table 25**.

Injectable Anesthesia

Over many years, the inhalation anesthetics have shown many advantages compared to injectable anesthetics. However, some very useful infusion anesthetics from human medicine have been adopted for swine. In general, infusion anesthetics have to be dosed intravenously, and a catheter is needed (in most cases in the ear vein). If swine have to be transported, it is necessary to have safe access, and a central catheter is recommended. In general, recovery time is long for most injectable anesthetics. Furthermore, it is not simple to change the anesthetic stage under injectable anesthesia. One of the preferred infusion anesthetics is propofol, although a range of useful drugs exist.[102,103] Some drugs for injectable anesthesia are presented in **Table 26**.

Propofol

Propofol is a non-barbiturate that can be administrated intravenously only (**Figure 16**).[103] It is cleared from plasma at a high rate, and

TABLE 26 DRUGS AND DRUG COMBINATIONS USED FOR INJECTION AND INFUSION ANESTHESIA OF SWINE[102–105]

Drug	Route	Dose	Remarks
Propofol	IV	4–10 mg/kg/h	
Acepromazine	IM	0.5–1.1 mg/kg	Ketamine administered 15 min after acepromazine
Ketamine		15–33 mg/kg	
Ketamine	IM	10– 8 mg/kg	
Diazepam		1–2 mg/kg	
Ketamine	IM	15 mg/kg	
Xylazine		2 mg/kg	
Ketamine	IM	10 mg/kg	
Medetomidine		80 µg/kg	
Butorphanol		220 µg/kg	
Ketamine	IM	5 mg/kg	Neuroleptanalgesia evaluated in Göttingen minipigs
Xylazine		2 mg/kg	
Butorphanol		220 µg/kg	

IV = intravenous; IM = intramuscular

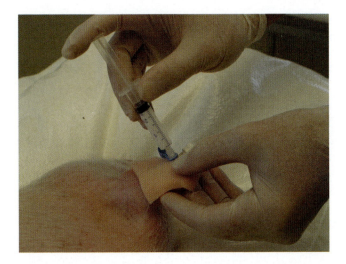

Figure 16 Induction of anesthesia with propofol, 1.5–5 mg/kg IV, through an over-the-needle catheter in the ear vein. (Photo courtesy of Ellegaard Göttingen Minipigs, Soroe Landevej, Dalmose.)

therefore consciousness returns more rapidly than with most other infusion anesthetics. As with isoflurane, propofol offers only minimal analgesia, so it is often combined with a strong analgesic, such as fentanyl. A major disadvantage for the use of propofol in swine is the high price; moreover, a bottle of propofol can be used only the first few hours after opening because of the lack of a bacterial preservative.

Ketamine and Other Dissociatives

Ketamine and other dissociatives produce a state of cataleptoid sedation with increased muscle tone and amnesia. Ketamine has been widely used to induce anesthesia in swine. However, its use as a mono-anesthetic in swine is not acceptable, since it does not induce effective anaesthesia.[113] Ketamine can be combined with drugs such as adreno-receptor (α2) agonists, benzodiazepines, and opiates. Combinations can be used for short- and long-lasting anesthesia of swine. Ketamine has a minimal depression effect on the cardiovascular system.

Barbiturates

Pentobarbitone and thiopental have no analgesic effects. Pentobarbitone causes severe cardiovascular depression and prolonged recovery. Swine are prone to apnea secondary to administration. Pentobarbitone should not be used as a sole agent. Nor should it be used in recovery studies, since recovery from pentobarbitone anesthesia is slow, with risk of hypothermia.

Adrenoreceptor (α2) Agonists

Xylazine is the most commonly used α2-adrenoceptor agonist in swine, though detomidine and medetomidine can also be used. While xylazine is a very potent sedative in ruminants, it is not in swine. Thus, in swine, xylazine must always be combined with other drugs, such as ketamine. The xylazine-ketamine mixture is popular for porcine anesthesia. Overdoses with an α2-adrenoreceptor agonist can be antagonized with yohimbine (0.2 mg/kg IV).

Benzodiazepines

Diazepam and midazolam are widely used for sedation and anesthesia of swine. As single drugs they only have a sedative effect, and are safe to use. However, they are often combined with ketamine or xylazine for the induction and maintenance of anesthesia. A mixture of ketamine and midazolam is useful for premedication (IM), induction (IV), and maintenance (IV) of anesthesia in swine used in scanning studies. Compared with their use in humans, benzodiazepines must be dosed in high concentrations in swine.

Alpha-Chloralose

Alpha-chloralose has been used in research for many years, and some researchers still use it to induce anesthesia in swine. However, today other anesthetics are preferred. Alpha-chloralose offers only minimal analgesia, the onset of anesthesia is slow, and spontaneous leg movements are often observed during anesthesia. Alpha-chloralose can be combined with N_2O, morphine, ketamine, or butorphanol.

Medetomidine-Butorphanol-Ketamine (MBK)

The drug mixture of ketamine, medetomidine, and butorphanol is useful for short-term anesthesia. It provides appropriate anesthesia and analgesia for surgical procedures for 30 to 45 minutes. Swine should be premedicated with atropine. The MBK mixture is administrated IM, and anesthesia is induced rapidly.

Supplementary Analgesia

For painful procedures, supplementary analgesics are needed. Opiates can be used as IV infusions during cardiac surgery. Because opiates do not decrease myocardial contractility and coronal blood flow, they

TABLE 27 SUPPLEMENTARY ANALGESIA FOR
PAINFUL SURGERY[103]

Drug	Infusion
Alfentanil	6 µg/kg/h
Fentanyl	30–100 µg/kg/h
Remifentanil	30–60 µg/kg/h
Sufentanil	15–30 µg/kg/h

are useful for cardiac surgery. However, anticholinergics have to be given to prevent the dose-related bradycardial effect of opiates. The dose of opiates has to be adjusted after effects are observed; some general doses are shown in **Table 27**.

Fluid Infusion

Fasting swine are often moderately dehydrated, and therefore IV fluid administration during anesthesia is recommended.[103] As a guideline, 5–10 mL/kg/h of warm saline can be administrated IV by a drop infusion. Insufficient urine production, extensive blood loss, and hypotension are indicators for higher infusion rates. However, flow rates that are too high may increase the risk of lung edema.

Depth of Anesthesia

It is important to assess the depth of the anesthesia continuously. During surgical anesthesia, the pig should not react to painful stimuli.[102] A rat-tooth forceps can be used to apply a stimulus in order to predict the absence of response to a surgical stimulus. A sharp pinch with a toothed forceps to the footpad will cause the foot to withdraw or a muscle twitch when the anesthesia level is insufficient. Stimulus to the nasal septum or the lateral lip fold, applied at moderate pressure, will result in head shaking, jaw clamping, or vocalization. Stimulus to the perianal region will be followed by contraction of the sphincter muscle. Autonomic responses such as increase in heart rate may indicate inadequate anesthetic depth.

Monitoring during Anesthesia

Several parameters can be monitored during the anesthesia of swine. **Table 28** shows the reference intervals for monitoring parameters in anesthetized swine. In general, the cardiovascular system, pulmonary system, and body temperature should be monitored.[103] Heart

TABLE 28 REFERENCE VALUES FOR THE MONITORING OF ANESTHETIZED SWINE

Organ System	Parameter	Göttingen Minipigs	Domestic Swine
Cardiovascular	Heart rate	94–116 BPM	68–98 BPM
	Mean blood pressure	93–111 mmHg	83–111 mmHg
	SaO2	95–100%	95–100%
Pulmonary	ETCO2	38–42 mmHg	37–43 mmHg
	PaCO2	38–42 mmHg	37–43 mmHg
	PaO2	68–74 mmHg	92–126 mmHg
	Respiration rate	17–23 RPM	11–29 RPM
Temperature	Rectal temperature	38–40°C	37–38°C

rate can be measured by electrocardiography (ECG). However, one should remember that ECG assesses only the electrical activity of the heart and not the mechanical activity. Blood pressure can be measured by invasive or non-invasive methods. If arterial access can be obtained, continuous invasive blood pressure monitoring is preferred. Non-invasive blood pressure monitoring can be easily obtained by placing a cuff around a leg. Using this method, blood pressure is measured only every 3 to 5 minutes. An alternative non-invasive method has been developed for domestic swine, where a neonatal blood pressure cuff is secured around the base of the tail. This method is a good substitute for invasive measurements of arterial blood pressure. Pulse oximetry measures the percentage of peripheral oxygen saturation (SpO_2) of arterial blood. The instrument also measures pulse. Suitable sites for placing the probe include the tongue, tail, and ears. Blood gases can be used to measure the lung gas exchange. Blood gas analyzers measure the partial arterial pressure of CO_2 ($PaCO_2$) and oxygen (PaO_2). End-tidal CO_2 ($ETCO_2$) can be used as an alternative for $PaCO_2$ measurements.

Neuromuscular Blocking Agents

Neuromuscular blocking drugs are used to produce paralysis, and as such can be used to stabilize mechanical ventilation by blocking spontaneous respiration and to prevent muscle movements during electrosurgery.[102] However, since the administration of neuromuscular blocking agents prevents movements in response to pain, they must be used with great care. When using neuromuscular blocking agents, monitoring of anesthetic depth is crucial; changes in heart

TABLE 29 NEUROMUSCULAR BLOCKING DRUGS
USED IN SWINE[102]

Alcuronium	0.25 mg/kg IV
Gallamine	2 mg/kg IV
Pancuronium	0.06 mg/kg IV
Suxamethonium	2 mg/kg IV
Vecuronium	0.15 mg/kg IV

rate or blood pressure are indicators of pain. Some commonly used neuromuscular drugs are shown in **Table 29**.

Postoperative Management

Recovery after anesthesia should always take place in a recovery box where the animal cannot injure itself, a box with smooth walls and ample bedding material. The room temperature must be 20–25°C to prevent hypothermia, or alternatively, a circulating water heating blanket should be used. An infrared heating lamp should be used only if the risk of thermal injury can be avoided by a thermostatic control. If N_2O has been used for inhalation anesthesia, the pig must be ventilated with pure oxygen for at least 10 minutes to prevent hypoxia. Until extubation and the recovery of the righting reflex, pulse, temperature, and respiration rate must be monitored at least every 5 minutes. Extubation should be performed only when a strong swallowing reflex is apparent. If it is done too early, a high risk of hypoxia is present, and if done too late, there is a risk of biting the tube. Following recovery, the pig has to be monitored daily until removal of the sutures. Antibiotics may be applied during recovery from surgery and the first 1 to 3 days thereafter. However, animal studies have shown that antibiotics administered intravenously one hour prior to surgery are much more effective than a first treatment with antibiotics after surgery. In case of postoperative vomiting, the respiration passages should be cleared. Once the pig is fully conscious, food and water are given. As swine usually have been fasted several hours prior to the surgery, they should show a good appetite.

Clinical Signs of Postoperative Pain

Because no objective tools for pain assessment in swine exist, swine have to be clinically examined several times during the first 1 to 3 days after surgery, as well as the first day after withdrawal of analgesic administration. Since the animal care staff is responsible for

TABLE 30 PAIN SCORE SYSTEM FOR SWINE BASED ON PALPATION OF THE SURGICAL SITE

Score	Description
1	Deep palpation of the surgical site and immediate surrounding tissue* does not provoke a response.
2	Deep palpation of the surgical site and immediate surrounding tissue provokes a response, but a similar response can be seen on the contralateral side of the limb, suggesting a hyperesthetic or hyperreflective state.
3	Deep palpation of the surgical site and immediate surrounding tissue provokes a response much greater than a stimulus on a nonsurgical part of the body. Probably indicative of some pain, and appropriate analgesia should be administered.
4	Deep palpation of the surgical site and immediate surrounding tissue provokes a response much greater than a stimulus on a nonsurgical part of the body and is accompanied by vocalization in an otherwise quiet patient. Requires analgesia.

* Remember sterility, and use a gloved hand.
Source: Data from Swindle, M.M. 2007. *Swine in the Laboratory: Surgery, Anesthesia, Imaging, and Experimental Technique.* Boca Raton: CRC Press.

daily care, they often are the first to observe signs of pain, including behavioral changes such as lack of appetite, and a decreased interest in surroundings, and decreased activity overall. Note that opiates can induce nausea and anorexia in swine. Unresponsive sterna recumbence often indicates pain. Palpation may provoke pain responses that can be quantified in a pain score system (**Table 30**), which can be helpful in evaluating the need for analgesics. It is important that the swine recover rapidly after surgery. Pain control will positively influence wound healing.

Postoperative Analgesics

The goal of pain control is to interrupt the nociceptive process between the peripheral nociceptor and the cerebral cortex, and this is done by using balanced analgesia. Surgically induced pain may require the use of opiates for 1 to 2 days, or even 3 days, after major surgery. Nonsteroidal anti-inflammatory drugs (NSAID) can be used in combination with opiates 1 to1½ days postoperatively and can be continued for 1 to 1½ days after opiates are no longer administered. Local anesthetics (e.g., bupivacaine) can also be used. In general, opiates have stronger analgesic effects than NSAIDs, but they have to be administered very often.[109] Transdermal fentanyl patches have been tried, but titration of the dosage is difficult, and overdosing has been observed when swine ingested the patches. Butorphanol and buprenorphine are long-acting in swine, and may be the opiates of choice for post-

TABLE 31 POSTOPERATIVE ANALGESIA FOR SWINE

Drug Class	Drug	Dose	Interval	Route
Opiate	Buprenorphine	5–20 µg/kg	6–12 hours	IM
	Butorphanol	0.1–0.3 mg/kg	4–6 hours	IM SC
	Morphine	0.2–1.0 mg/kg	4 hours	IM
	Pethidine	2 mg/kg	2–4 hours	IM
NSAID	Acetylsalicylic acid	10–20 mg/kg	4 hours	PO
	Carprofen	1–4 mg/kg	daily	SC
	Flunixin	1–2 mg/kg	daily	SC
	Ketoprofen	3 mg/kg	daily	IM
Local	Bupivacaine		10–12 hours	Infiltration

IM = intramuscular; SC = subcutaneous; PO = oral
Source: Data from Flecknell, P. 1996. *Laboratory Animal Anaesthesia.* London: Academic Press.

operative analgesia. Butorphanol is a potent agonist-antagonist that has three to five times the potency of morphine. As an antagonist, it can be used to reverse the action of fentanyl while maintaining some analgesic effects by its action at kappa-receptors. Buprenorphine is a partial mu agonist. Most NSAIDs have only to be dosed daily, and thus are very popular for postoperative analgesia. Unlike opiates, NSAIDs are effective only against weaker pain. They are also useful against inflammation-induced pain. They may be combined with opiates for the treatment of acute pain. Because of the anti-inflammatory effect of NSAIDs, they very often affect porcine research models. Carprofen has good analgesic effects for orthopedic and soft-tissue pain. It has minimal effects on gastrointestinal bleeding and nephrotoxicity, and it can be administrated preoperatively. Flunixin has a prolonged effect, but should not be administered for longer than three days. Ketoprofen can be used for postoperative and chronic pain. For both opiates and NSAIDs, side effects are observed only after several days. As an alternative to opiates and NSAIDs, local anesthetics infiltrated at the site of incision will provide long-term analgesia (10–12 hours) without side effects and anorexia, but administration is more difficult. Preemptive analgesia refers to the application of balanced analgesia prior to exposing the patient to noxious stimuli. Doses of opiates and NSAIDs are shown in **Table 31**.

Anesthetic Emergencies

Insufficient ventilation of swine can lead to respiratory arrest and death. If there are any signs of respiratory arrest, the oxygen supply should be checked. Administration of any anesthetics must be stopped,

TABLE 32 SOME EMERGENCY DRUGS FOR SWINE DURING ANESTHESIA

Drug	Dose	Indication
Adrenaline	0.02 mg/kg IV	Cardiac arrest
Atropine	0.05 mg/kg IV	Bradycardia
Dantrolene	3.5–5 mg/kg IV	Malignant hyperthermia
Digoxin	0.01–0.04 mg/kg IV	Supraventricular arrhythmias
Dopamine	2–20 µg/kg/min IV	Hypotension
Doxapram	5–10 mg/kg IV	Hypoventilation
Lidocaine	2–4 mg/kg + 50 µg/kg/min IV	Arrhythmias
Pentobarbitone	100–150 mg/kg IV	Euthanasia
Propranolol	0.04–0.06 mg/kg IV	Tachycardia
Succinylcholine	1–2 mg/kg IV	Laryngospasm

Source: Data from Swindle, M.M. 2007. *Swine in the Laboratory: Surgery, Anesthesia, Imaging, and Experimental Technique.* Boca Raton: CRC Press.

and the pig should be mechanically ventilated with pure oxygen. If the pig is not intubated, airways must be checked to see whether the tongue is obstructing the larynx. If necessary, the respiratory tract should be aspirated to remove blood, vomitus, and bronchial secretions. Respiration can be stimulated by injection of doxapram, repeated every 15 to 20 minutes if needed. The cardiovascular system is depressed when overdosed with anesthetics. In case of cardiovascular failure, administration of anesthetics should be stopped and ventilation assisted with pure oxygen. Rapid IV infusion of saline or plasma expanders is useful. In case of cardiac arrest, external cardiac massage (80–100 compressions per minute) should be given. Lateral rather than sterna compression should be used in swine. Adrenaline can be administrated intracardially, IV, or into the endotracheal tube. Malignant hyperthermia, also known as porcine stress syndrome, is another serious anesthetic emergency. This condition is a hereditary disease known in many breeds of swine, though not in minipigs. The symptoms begin shortly after anesthesia with halothane, and consist of tachycardia, stiffness of muscles, tachypnea, and hyperventilation. After a short period, the condition progresses to apnea, cyanosis of the skin, and rapid rise in body temperature. At this point, it is best to turn off the halothane and use another kind of anesthesia.[103] The drug dantrolene can be administered in this situation. This and other emergency drugs are presented in **Table 32**.

Laryngospasm and Edema

Insufficient anesthesia level, excessive manipulation, unsuccessful attempts to intubate, and extubation may lead to laryngospasm. It can be treated with succinylcholine and immediate ventilation with oxygen. Spraying a small volume of Xylocaine prior to intubation will decrease this risk. Overinfiltration of the cuff can cause swelling and edema of the airways. The tissue trauma becomes apparent after extubation. It can be prevented and treated with corticosteroids, diuretics, or NSAIDs.

euthanasia

Swine can be euthanized humanely by an intravenous overdose of 20% solution of sodium pentobarbitone (100–150 mg/kg IV). Pentobarbitone can be administrated directly into a central vein or into an ear vein after sedation. Neonatal animals can be euthanized IP with pentobarbitone, though this drug may enlarge the size of the spleen and other organs.[103] As an alternative to pentobarbitone, potassium chloride (KCl) can be injected (2 mmol/kg IV), or a captive bolt can be used (**Figure 17**). Note that KCl may be used only in deeply anesthetized swine, and a captive bolt must always be followed by exsanguination. At slaughterhouses, CO_2 at concentrations over 70% is used for stunning. This causes asphyxiation and does not leave any drug residues in meat products, but this method is rarely used for laboratory swine. With any method for euthanasia, cessation of respiration and heartbeat and loss of reflexes are good indicators of death.

Figure 17 Captive-bolt pistol. The bolt pistol is placed on the cross point of two imaginary lines from the ears to the eyes. (Photo courtesy of Ellegaard Göttingen Minipigs, Soroe Landevej, Dalmose.)

experimental techniques

Swine have been used as experimental animals throughout history, mainly in anatomical and physiological research. In his *De Humani Corporis Fabrica*, from 1543, Vesalius referred to swine experiments performed by the Roman physician Galen (ad 130–200), but he made observations about swine himself as well. Also Harvey, in his *Anatomica de Motu Cordis et Sanguinis Animalibus* (1628), and Bernard, in his *Introduction à l'étude de la médecine expérimentale* (1865), described observations made of swine. At present, swine are used in a wide range of disciplines.

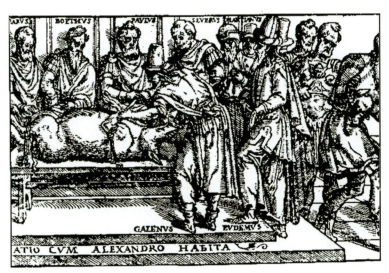

Figure 18 Illustration of Galen from *De Humani Corporis Fabrica* (1543). (Photo courtesy of Ellegaard Göttingen Minipigs, Soroe Landevej, Dalmose.)

restraint

Restraint techniques in swine fall into three categories: manual restraint, mechanical restraint, and chemical restraint.[101–104] An often forgotten but simple method of restraint of placid sows is to stroke the abdomen while talking softly to the animal. This will allow general examination. In any case, the animals have to be approached first.

Even though domestic swine and most minipigs have a gentle character, defensive behavior can occur, especially in sows with piglets and in boars. Raised bristles and tail and head shaking may indicate an upcoming attack. Both sows and boars have tusks, which can cause deep wounds, and swine are strong animals. Swine should always be approached carefully.

Swine have to be approached quietly, and talking softly will calm the animal. Crouching is less threatening to the animal than standing bent over it. Usually, the swine will face the approaching person in an awaiting attitude. When swine feel threatened, they face toward the source of threat to avoid being attacked from the side, since the natural offensive behavior of swine is butting and biting on the shoulder and neck region. Young swine and miniature swine may be seized by a front leg in this situation, and lifted up on to the arms. If the animal is facing away or tries to escape, a hind leg may be grabbed instead. The tail and ears should not be used to lift the animal.

Larger swine should preferably be trained to walk from their pen, and they can be driven in the right direction by walking behind them. A board can be used to drive the animal, and can also be used to restrain large swine against a wall.

Young swine and miniature swine under 15–20 kg (33–44 lb) can be lifted and then manually restrained by holding them in one's arms. One arm is positioned under the abdomen or hind quarters, while the other arm supports the bow. This method is suitable for a single IM administration in the cervical muscles or for a short examination. For examining the abdomen or feet, a swine can be positioned on its back in the lap of a sitting person. Alternatively, the animal can be held by both hind legs, head down.

When blood sampling is needed, drugs are administered. When a short examination is necessary, large swine are often restrained with a nose snare. A loop, usually a steel cable, is placed around the upper jaw behind the tusks to hold the animal. In large swine, this is often the only method of restraint aside from chemical restraint. It is a stressful method and should be avoided in young and miniature swine.

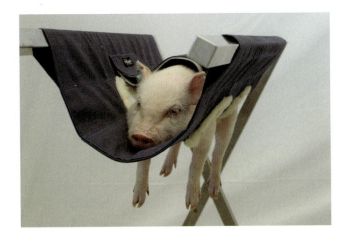

Figure 19 Restraint of a juvenile minipig in a hammock. (Photo courtesy of Ellegaard Göttingen Minipigs, Soroe Landevej, Dalmose.)

Swine under 50 kg (110 lb) can be restrained in dorsal recumbency in a V-shaped trough, or on the abdomen in a hammock.[117] Both methods are used mainly for short restraint for blood sampling, but trained, conscious swine may be restrained in a hammock over longer periods, for infusion studies or ECG recording, for example. In contrast to dogs being restrained in hammocks, swine should hang freely, without touching the floor, as in **Figure 19**, since swine cannot be trained to stand still. Mechanical restraint methods for specific purposes may be developed individually, such as restraint in a bucket for intranasal inoculation or swabbing.[115]

Inactivation of swine by sedatives is called chemical restraint. The most common sedative is azaperone (Stresnil, Sedaperone), which should be dosed at 4 mg/kg IM (**Table 19**). A single IM administration in large swine can be carried out in the animal's pen by using a syringe with an infusion line. The hypodermic needle is inserted into the cervical muscles, and the syringe emptied. The extension tube allows some degree of movement of the animal (see Chapter 4).

sampling techniques

Blood Sampling

There are few percutaneous access sites to veins in swine, since veins and arteries lie either under thick skin and subcutaneous fat, or deep in between muscles. Possible sampling sites are the[102,103]

- cranial caval vein
- jugular vein
- femoral vein
- saphenous vein
- cephalic vein
- carotid artery
- femoral artery
- cardiac puncture

The cranial caval vein and femoral vein are most often used, and the procedure will be described. The ear veins are not suited for withdrawal of large blood samples since these veins, being too small, will collapse upon withdrawal of blood. Arteries should be avoided for blood sampling, since puncture may lead to the formation of hematoma.

The best location for blood sampling is the cranial caval vein (see **Figure 20**), with the swine in dorsal recumbency. Large swine have to be restrained standing, with a snare. For dorsal recumbency, the swine may be placed on a tabletop, or in a V-shaped trough, the latter being preferable as it limits the movements of the animal. Along with the person taking the blood sample, one or two assistants are required

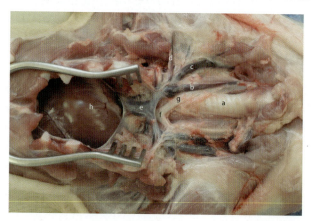

Figure 20 Anatomy of the brachiocephalic plexus, which can be accessed by inserting a long hypodermic needle cranial to the sternum at an angle of under 45 degrees. A too steep insertion of the needle will puncture the carotid trunk. a) trachea; b) internal jugular vein; c) external jugular vein; d) cervical vein; e) cranial caval vein; f) carotid trunk; g) common carotid artery; h) heart. (Photo courtesy of Ellegaard Göttingen Minipigs, Soroe Landevej, Dalmose.)

to restrain the animal. The front legs are bent caudally, and the neck is stretched by holding the animal by its snout. A second person may hold the hind legs. To prevent a righting reflex when placing the animal on its back, the animal should not be rolled over the longitudinal axis into the V-shaped trough, but over the transversal axis. When this is done in a quiet manner, the animal will be relaxed and easy to restrain. The manubrium of the sternum is now palpated, and a 21-gauge hypodermic needle inserted into the medial plane at an angle of 45–60° to an imaginary horizontal line, a few millimeters cranial to the manubrium of the sternum. Creating a slight vacuum in the syringe while inserting the needle will facilitate easy detection when a vein is punctured, since blood will stream into the syringe. A vacutainer blood-sampling system will ease the procedure. For swine up to 10–15 kg (22–33 lb), a hypodermic needle of 25 mm (1 inch) can be used. Larger swine, from 15 to 50 kg (33 to 110 lb), require a needle of 45 mm (1¾ inches), whereas swine over 50 kg (110 lb) require a needle size of 18 gauge/65 mm (2½ inches). The hypodermic needle should not be inserted at too flat an angle, under 45°, because of the risk of puncturing the pericardium or the heart, nor too steep, because of the risk of puncturing the carotid trunk.

Alternatively, a hypodermic needle can be inserted into the jugular furrow. The point of the needle should be directed towards the medial plane, dorsal to the manubrium of the sternum. The angle to an imaginary horizontal line should be 45°. The two methods described will puncture the brachiocephalic plexus, cranial to the cranial caval vein. The caval vein could also become punctured with these two methods. When inserting a needle into the jugular furrow in the sagittal plane, the jugular vein should be punctured. However, in young swine and miniature swine, this method is not as successful as puncturing the brachiocephalic plexus.

The blood volume of a single sample should not exceed 10% of the total blood volume.[116] Since swine have a blood volume of 75 mL/kg, a maximum volume of 7.5 mL/kg can be taken. An interval of 2 weeks between single samples should be respected. In repeated sampling, as used in pharmacokinetic studies, it is recommended not to take more than a total volume of 7.5 mL/kg. For a 10-kg swine, this corresponds to 15 samples of 5 mL within 24 hours, whereafter the animal is not sampled for 2 weeks. Unlimited access to drinking water should be secured to allow the animal to rehydrate itself. The site of sampling should be varied in repeated sampling to reduce the risk of hematoma. For accessing the brachiocephalic plexus, sampling could be varied between the right and left jugular fossa

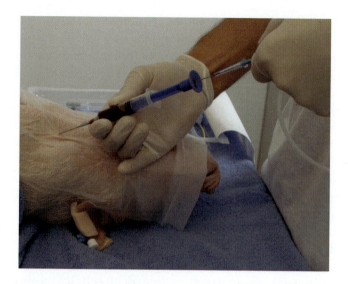

Figure 21 The brachiocephalic plexus can be used to place a per-cutaneous catheter for sampling and infusion over several days. As soon as the plexus is punctured and venal blood is flowing into the syringe, a guide wire is inserted through the hypodermic needle. In some kits, this can be done unbloodily by inserting the guide wire through the plunger of the syringe. (Photo courtesy of Ellegaard Göttingen Minipigs, Soroe Landevej, Dalmose.)

and the medial plane, or, for sampling over several days, a catheter can be placed percutaneously (**Figures 21** and **22**.)

If repeated sampling is necessary over a long period, surgical catheterization may be considered.[101,117,118] Methods are described below.

Urine and Feces

Only in sows can the urethra be catheterized, since boars have a corkscrewed penis tip and a sigmoid curve in the body of the penis. However, urine sampling in swine by catheterization of the urethra is seldom performed, since it requires an anesthetized animal. A large suburethral diverticulum precludes catheterization in conscious swine, since catheterization is difficult without the help of a speculum.

Urine and feces are often collected by means of a metabolic cage. Cages designed for dogs may be used for small or miniature swine. Feces will stay behind on the grid floor of the metabolic cage, while urine is collected via a pan with a funnel. Contamination of urine with feces is unavoidable, but can be minimized by collecting the feces regularly.

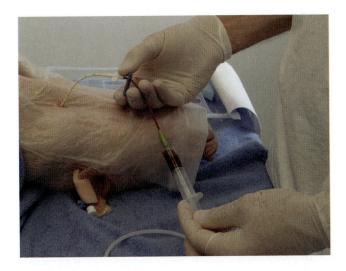

Figure 22 After the catheter is placed, the patency is tested by aspirating some blood. Then the catheter is flushed with heparinized saline, closed, and draped with adhesive tape. A jacket can be used for protection or for carrying a portable infusion pump. (Photo courtesy of Ellegaard Göttingen Minipigs, Soroe Landevej, Dalmose.)

Cerebrospinal Fluid

Cerebrospinal fluid can be collected from the anesthetized swine by puncturing the cisterna magna. Sampling cerebrospinal fluid in conscious swine is not possible, because of the non-cooperative behavior of swine. The swine should be placed in lateral recumbency, and the head fully flexed. The cervical area should be clipped and disinfected, with 1% chlorhexidine in 70% ethanol, for example. A 20-gauge spinal needle with a length of 90 mm (3½ inches) and provided with a stylet should be inserted about 5 cm (2 inches) caudal to the occipital protuberance, and advanced slowly towards the mouth. The needle should be kept parallel to the table surface. The atlanto-occipital membrane and dura mater of the cisterna magna lie about 7 cm (2¾ inches) deep in a 40-kg (88-lb) swine. A volume of 5–10 mL can be collected at intervals of 4 days.[119]

Bile and Pancreatic Excretions

Bile and pancreatic excretions can be collected in a chronically catheterized animal.[101,120] Methods are described later in this chapter.

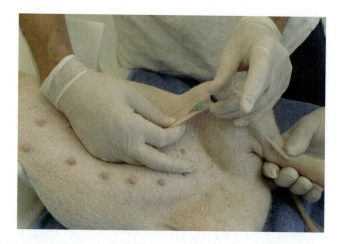

Figure 23 SC administration in the flank. (Photo courtesy of Ellegaard Göttingen Minipigs, Soroe Landevej, Dalmose.)

administration of compounds

Small needle sizes (19–21 gauge) can be used for most swine for SC, IM, and IV injections.[101]

Subcutaneous administration (SC) is not easily executed, since the swine is a fixed-skin animal, as are humans. Small volumes may be given on the lateral side of the neck immediately caudal to the ear, or in the flank[101,103] (**Figure 23**). In obese swine, SC injections are often inadvertently administered into the subcutaneous fat.

Intramuscular administration (IM) is preferably given in the cervical muscle groups, in the region overlaid by the ear, close to the transition from hairy to hairless skin (**Figure 24**). This region has the thinnest layer of fat, with the muscles closest to the skin. The semimembranosus, semitendinosus, and gluteal muscles in the hind leg may also be used (**Figure 25**), but cause more resentment to injection than the neck region. The drugs are administered through a 19- to 21-gauge needle, with a minimum length of 3–4 cm (1⅛–1¾ inches) to ensure that the injection is given into the muscle. The needle can be connected to an extension tube, which will allow the swine to move freely while the drug is being injected.

Intravenous administration (IV) often requires a sedated animal. The central or ventrolateral ear veins are most commonly used (**Figure 26**). The artery is located on the dorsolateral aspect of the ear.[103] A 20- to 22-gauge hypodermic needle can be used. The use of a butterfly needle has the advantage that the needle can be taped to

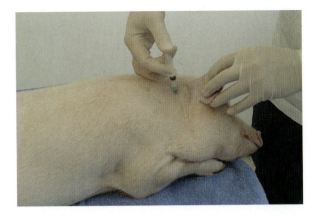

Figure 24 IM administration in the neck. (Photo courtesy of Ellegaard Göttingen Minipigs, Soroe Landevej, Dalmose.)

Figure 25 IM administration in the hind leg. (Photo courtesy of Ellegaard Göttingen Minipigs, Soroe Landevej, Dalmose.)

the ear. When a Venflon® over-the-needle cannula or similar cannula is used, the cannula can be kept in the ear vein for 2 to 3 days. The use of a stylet will prevent coagulation inside the cannula. For an emergency situation, the cranial caval vein can be used. This vein can be accessed as previously described in this chapter.

Intraperitoneal administration (IP) is seldom used in swine.

basic surgical procedures[101,103]

All surgery should be performed using aseptic procedures and techniques. Anything that comes into contact with the surgical site should

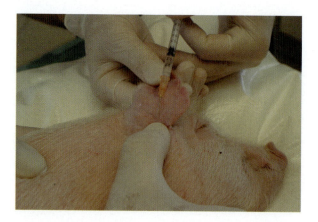

Figure 26 IV administration in the ear vein. (Photo courtesy of Ellegaard Göttingen Minipigs, Soroe Landevej, Dalmose.)

be sterile, including instruments, suture materials, syringes, and needles. To prepare the animal for surgery, the hair should be removed from the surgical site with a clipper, and the skin scrubbed with soap and water and disinfected with alcohol and an iodine solution. The surgical site should be isolated with sterile drapes. The surgeon and assisting staff should wear caps, facemasks, sterile gowns, and sterile surgical gloves.

Catheterization

The Carotid Artery and the Internal Jugular Vein

A 5-cm (2-inch) longitudinal incision is made in the midline approximately 3 cm (1⅛ inch) cranial to the manubrium of the sternum. A longitudinal incision is made in the sternohyoideus muscle down to the trachea. The vessels are located on either side of the trachea. The carotid artery and the jugular vein are isolated for a length of 2 cm (¾ inch) using blunt dissection. There should be as little manipulation as possible of the vein since it contracts easily, which would make catheterization more difficult. After isolation of the vessel, a length of suture material is placed around the vessel at the distal and the proximal ends. The distal suture is tied. The proximal suture is placed loosely around the vessel.

The Carotid Artery

An arterial clamp is positioned on the artery just beyond the proximal suture. A small cut approximately ⅓ of the diameter of the arterial wall is made with a pair of fine scissors (preferably micro-scissors) close to the distal ligature. A catheter is placed in the artery, using a vascular guide to hold the lumen of the artery open, and while the arterial clamp is slowly loosened, the catheter is introduced into the artery. The proximal suture is tied around the catheter and the artery, and the catheter is fixed using the distal ligature ends.

The Internal Jugular Vein

A catheter is placed in the vein in the same manner as in the artery, except that it is not necessary to use the clamp. The proximal suture is lifted slightly to empty the vein, and the catheter is introduced into the vein using a vascular guide. It might be helpful to loosen the proximal suture a little just prior to entering the vascular lumen in order to fill the vein with blood.

The External Jugular Vein

A 5-cm (2-inch) longitudinal incision is made approximately 5 cm (2 inches) lateral to the midline in the same area as for the carotid artery. The vein is located deep between the brachiocephalic and the sternocephalic (sternomastoideus) muscles. The vein is cannulated in the same manner as described for the internal jugular vein.

The Cephalic Vein

The cephalic vein joins the external jugular vein and lies superficially in the axillary furrow. A 3-cm (1⅛ inch-) incision is made in the furrow, perpendicular to the midline and beginning 4 cm (1¾ inches) from the manubrium of the sternum. The vessel is dissected bluntly for a length of 2 cm (¾ inch). It is cannulated in the same manner as above.

The Femoral Artery and Vein

The femoral canal can be located by palpating the pulse in the fold between the sartorius and gracilis muscles. An incision is made starting cranial to the origin of the pectineal muscle and extending along the femoral canal distally on the leg. The subcutaneous tissue is dissected, and the two muscle groups are located and separated bluntly. The two vessels are located deep, right next to the saphenous nerve, which lies cranial to the artery. The vessels are carefully separated from the nerve, and then dissected over 2–3 cm (¾–1⅛ inches). Beware of the side branches of both the artery and the vein. These must be carefully ligated and transected. Both vessels are cannulated as described above.

Other vessels, such as the subcutaneous mammary veins and the median coccygeal artery, are also suitable for cannulation.

Laparotomy

A 15-cm (6-inch) midline incision is made from the xiphoid process of the sternum, through the subcutaneous layer of the skin and the muscle layer in the linea alba. The peritoneum is grasped with a tissue forceps, lifted, and incised with a small hole using a scalpel. The blunt tip of a pair of scissors is inserted and the peritoneum is incised, lifting the tissue to make sure not to incise the underlying organs.

Portal Vein Cannulation

The spleen is exposed, and the gastrosplenic veins are located between the large curve of the stomach and the spleen. The vein is dissected carefully, without traumatizing the spleen tissue. Two sutures are placed around the vessel. The distal suture is tied. A small cut is made in the vein, and a soft silicone tube is placed in the vein and secured very loosely with the proximal suture. Too tight a ligature will prevent the further insertion of the catheter. The tube is carefully directed into the splenic vein (sharp bend), and further introduced 30–35 cm (12–14 inches) until it can be palpated in the portal vein. The catheter has a tendency to turn downwards into the gastroduodenal vein. To prevent this, the tube should be palpated by putting your fingers into the epiploic foramen. Here you should be able to guide the tube gently in the cranial direction.

The Hepatic Vein

The left lobe of the liver is identified. An intestinal clamp is put on the lobe to control bleeding, and a transverse incision is made a few centimeters from the edge. A central vein is identified, and a silicone rubber tube is introduced approximately 8 cm (3⅛ inches) into the liver after removal of the clamp. The wound is closed by putting a hemostatic material into the wound to control bleeding and then suturing the wound.

The Pancreatic Duct

The pancreatic duct is identified near the proximal end of the pancreas, close to the duodenum. The peritoneum is transected, and the duct is dissected. Two ligatures are placed, and the one next to the duodenum is ligated. The duct is cut with micro-scissors, and the catheter introduced using a vascular guide.

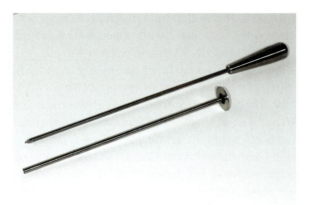

Figure 27 Trocard for subcutaneous tunneling of the free catheter end. (Photo courtesy of Ellegaard Göttingen Minipigs, Soroe Landevej, Dalmose.)

Exteriorization of Catheters

In acute experiments, the wound may be closed by single sutures or wound clamps, leaving the catheter end exposed for sampling, pressure measurements, or other procedures. When the animal needs to recover from anesthesia, however, further surgical preparation of the exteriorization site of the catheter is necessary. Often this requires subcutaneous tunneling of the catheter toward a site that is easily accessible. Free catheter ends are often tunneled to the midline of the back, between the shoulders, but when the catheter has to be connected to a vascular access port, often a site over the dorsal part of the scapula is chosen. This ensures that the vascular access port is placed over a firm underground, facilitating fixation and access. When tunneling a catheter to a site, sufficient free length of the catheter has to be taken into account before implantation. For subcutaneous tunneling, a long trocard (**Figure 27**) is needed.

Careful preparation of the exteriorization site is needed, and the skin needs to be cleaned and disinfected thoroughly. For a long patency of the catheter, it is of the utmost importance to work under sterile conditions, and great attention should be paid to wound closure and catheter maintenance.

Catheter Sizes and Material

No general guidelines on catheter diameter and length can be given for swine, since catheter size is dependent on age and breed. Thus, choosing a suitable catheter depends much on experience. Generally, a catheter diameter of 2 to 3 mm can be used in the majority of vessels. It is a good idea to select a catheter with a surplus of length. It is always possible to shorten a catheter according to individual anatomical markers during implantation, whereas nothing can be done when a catheter is too short. When implanting catheters in young, growing swine, take an additional length into account to compensate for growth. The extra length of catheter can be placed as a loop in a subcutaneous pouch near the implantation site. Catheters made of silicone and polyurethane are suitable for chronic implantation, whereas catheters made of polyethylene are recommended only for acute experiments.

Catheter Maintenance

Catheters need to be flushed daily using sterile heparinized saline (2–5 U/mL). If an animal is receiving a continuous IV infusion, flushing is not necessary during the infusion period. For each injection or sampling procedure, sterility is necessary, and the skin and injection cap should be cleaned with isopropyl alcohol before each use to decrease the risk of introducing bacteria into the catheter site. The catheter exteriorization site should be cleaned daily with an antibacterial solution or ointment. Catheter-related sepsis is the most common cause of loss of patency of a chronically implanted catheter.

---

safety testing of chemicals and drugs

Miniature swine are increasingly used as the non-rodent species in safety testing of chemicals and drugs. Large swine are used to a lesser extent (with the exception of toxicity testing of agro-chemicals or other compounds available in large quantities) because body weight, and thus higher dosing, of large swine is the limiting factor. Miniature swine are used in all fields of safety testing, including dermal toxicology[121,122] and acute and chronic systemic toxicology by oral dosing or continuous infusion.[123,124] But miniature swine have also been used in such specialized fields as immunotoxicology,[123] reproductive toxicology,[115] and teratology.[126]

Dermal Toxicology

The procedures for acute dermal toxicity and acute dermal irritation are described in guidelines issued by the Organization for Economic

Cooperation and Development (OECD).[127] The animals should be prepared as follows:

- The back and sides of the swine are clipped one day in advance.
- The back is divided into 6 or 8 squares on both sides of the spine. The total surface of the squares should correspond to 10% of the total body surface.
- The total body surface can be estimated using:

$$BS = (70 \times BW^{0.75})/1000$$

$$BS = body\ surface\ (m^2) \quad BW = body\ weight\ (kg)$$

- The equation is based on the fact that animals have a basal metabolism of 70 kcal/kg and a heat production of 1000 kcal/m^2. The use of metabolic body weight (BW$^{0.75}$) makes the equation applicable for interspecies comparison.
- A prescribed amount of test compound is applied to the skin, so that all squares are covered, and dressed with gauze pads. The pads are kept in place with adhesive tape or covered by a larger gauze pad, kept in place with tubular net bandage.
- The duration of exposure should be in accordance with the relevant regulatory guideline. After the period of exposure, the pads are removed, and the compound is washed off the skin. The skin is examined for erythema and edema.[126]

Acute and Chronic Systemic Toxicology

Guidelines for acute and chronic systemic toxicology relevant for swine are described by OECD[127] and others. Guideline 409 (sub-chronic oral toxicity, non-rodent: 90 days) mentions miniature swine explicitly as an acceptable species.[128]

Oral administration of the test compound can be done by

- mixing the compound directly in the feed
- administering the compound in capsules into the oropharynx, via a balling gun
- administering the compound directly into the stomach, via gastric intubation

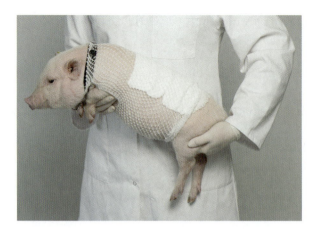

Figure 28 Dressing used during dermal exposure; the gauze pad is kept in place with a tubular net bandage. (Photo courtesy of Ellegaard Göttingen Minipigs, Soroe Landevej, Dalmose.)

Gastric intubation is the most controlled method of oral administration. Although swine are ready eaters, feed may have a reduced palatability after the compound has been mixed into it. Capsule dosing has the uncertainty of capsules not being swallowed. Swine have to be restrained for gastric intubation. The procedure is as follows:

- A sitting technician holds the swine in an upright position with its hindquarters held firmly between his or her legs.

- The front legs of the swine are bent caudally and held with one hand. The other hand stretches the neck of the swine, and the animal is pressed against the technician's chest. A seat fixed to the technician's chair, in which the swine can sit, will ease the restraint.

- A soft gastric tube 3–4 mm in diameter can be used for miniature swine of 5–15 kg (11–33 lb). For larger swine, a 6-mm gastric tube can be used. The length of the tube should reach from the mouth to past the last rib of the restrained animal.

- The tube is inserted into the mouth and introduced. Normally, the swine will protest and vocalize loudly. When the swine stops vocalizing, it should be checked whether the trachea has been intubated instead of the esophagus by listening to the respiration. The tube may be wetted with water before intubation, but for small diameters this is usually not necessary. The compound is injected into the stomach when the

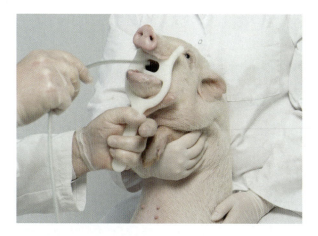

Figure 29 Oral administration by gastric intubation (gavaging). (Photo courtesy of Ellegaard Göttingen Minipigs, Soroe Landevej, Dalmose.)

tube is in place (**Figure 29**). No guidelines for maximum volumes are available, but a typical volume is 10–20 mL.

Intravenous administration of a test compound can be by single injection, repeated administration, or continuous infusion. Repeated administration and continuous infusion require a surgically prepared animal. Methods for single IV administration and for catheter implantation are described earlier in this chapter. A vascular access port can be used for repeated administration, as well as for blood sampling.[117,129] For continuous infusion, the catheter should be connected to an ambulatory infusion pump, which is placed in the pocket of a jacket on the back of the swine. The procedure is as follows:

- After the animal has been catheterized, the catheter is connected to the infusion pump. The catheter should be exteriorized on the shoulder, at the same level where the pocket of the jacket will be located. The reservoir of the infusion pump should be filled with saline.
- A well-fitting jacket is put on the swine while it is still anesthetized. The infusion pump is placed in a pocket of the jacket, and a counter weight is placed into the other pocket to balance the jacket.
- When the swine wakes from the anesthesia with the jacket in place, it will easily accept the jacket (**Figure 30**).

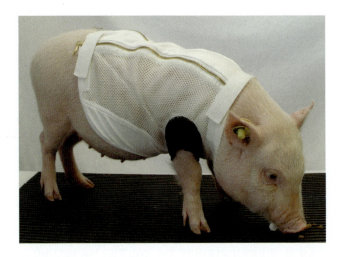

Figure 30 Minipig with a jacket for continuous infusion. (Photo courtesy of Ellegaard Göttingen Minipigs, Soroe Landevej, Dalmose.)

- The battery and reservoir should be changed daily. Seven days post-operatively the study can commence, and the saline is replaced with the test compound.

A typical repeated-dose study lasts for 28 days, whereas subchronic toxicity studies last for 90 days.

necropsy[95,96]

Necropsy is an important tool in the diagnosis of disease, but also in evaluating the pathological effects of an experimental procedure. Necropsy is an essential part in safety testing of chemicals and drugs. Whereas for the diagnosis of disease only affected organs are sampled for histopathological examination, in safety testing there is a narrowly described list of organs that are to be weighed and sampled. Organs that have to be examined are described by such regulatory bodies as the Organization for Economic Cooperation and Development and the Food and Drug Administration. **Table 33** gives an overview of organs to be sampled for histopathology.

Equipment

Necessary equipment includes rubber gloves and boots, a knife with a 5- to 7-inch blade, scalpel blades and holder, scissors, and forceps.

TABLE 33 ORGANS TO BE SAMPLED FOR HISTOPATHOLOGICAL EXAMINATION IN SAFETY TESTING OF CHEMICALS AND DRUGS

Organ/Tissue	FDA Redbook[a]	OECD[b]
Adrenals	left + right	left + right
Aorta	microscopy	microscopy
Bone marrow	microscopy	microscopy
Brain	three specified sections	three specified sections
Bronchi	microscopy	—
Cecum	microscopy	microscopy
Colon	microscopy	microscopy
Duodenum	microscopy	microscopy
Esophagus	microscopy	microscopy
Eyes	left + right	left + right
Femur	—	if signs of toxicity are present
Gall bladder	microscopy	microscopy
Heart	microscopy	microscopy
Ileum	microscopy	microscopy
Jejunum	microscopy	microscopy
Kidneys	left + right	left + right
Liver	microscopy	microscopy
Lungs	microscopy	microscopy
Lymph nodes	only one node	only one node
Mammary gland	only one gland	only one gland
Ovaries	left + right	left + right, including cervix
Pancreas	microscopy	microscopy
Pituitary	microscopy	microscopy
Prostate	microscopy	—
Rectum	microscopy	microscopy
Salivary glands	left	left + right
Sciatic nerves	left	left
Skeletal muscle	left	left
Skin	—	if signs of toxicity are present
Spinal cord	two specified sections	three specified sections
Spleen	microscopy	microscopy
Stomach	microscopy	microscopy
Testes	left + right	left + right
Thymus	microscopy	microscopy
Thyroid	including parathyroids	including parathyroids
Trachea	microscopy	microscopy
Urinary bladder	microscopy	microscopy
Uterus	microscopy	microscopy

[a] Food and Drug Administration, *Redbook*, 1982.
[b] Organization for Economic Cooperation and Development, *Guidelines for Testing of Chemicals*, Section 4: Health Effects, 1981.

To collect tissues and take samples for microbiological examination, containers with formalin, plastic bags, strings, and culture swabs are essential. Helpful equipment includes knife, rib cutters, microscope slides, labels, ruler, syringes, and hypodermic needles.

Preservation

The only necessary preservative is a buffered 4% formaldehyde solution (formalin). This can be prepared by adding 100 mL 40% formaldehyde to 900 mL water, and then adding 6.5 g diphasic anhydrous sodium phosphate and 4.0 g monobasic sodium phosphate. Specimens should not be thicker than 5 mm to allow proper preservation. The correct tissue-formalin ratio is 1:10. Specimens meant for microbiology should be kept in a plastic bag in a refrigerator; they should not be kept in formalin or frozen.

Procedure

Start by examining the external features, such as described for a clinical examination. If the animal is found dead, and the cornea is cloudy and the abdomen discolored green, a necropsy is often less valuable, since the animal has undergone autolysis.

For examination of internal structures, place the swine in left lateral recumbency, and then

- cut deep into the right axilla; extend the incision to the point of the mandible and to the anus (insert the knife into the subcutis and cut outward); reflect the skin; cut the right hip joint; open the stifle and reflect the limb; examine the inguinal lymph nodes.
- open the abdomen with a paracostal cut; extend to the pelvis with a paralumbar cut; reflect abdominal wall ventrally.
- incise the diaphragm and cut the ribs and remove them.
- examine the peritoneal and thoracic cavities; if exudates, adhesions, fibrin, or other pathological changes are present, collect tissue for culture and histopathology; blood and urine should be collected at this point.
- open the pericardium in situ, and examine.
- cut close to the bone on the medial side of both mandibles, and free the tongue; pull it caudally and cut the hyoid bones;

the soft tissue is cut along the vertebrae, and the cervical and thoracic viscera are freed intact.

- examine the oral cavity and the regional lymph nodes; incise the tongue, and open the esophagus.
- palpate the lungs and then open the trachea and major airways; the bronchial lymph nodes are cut, and finally the heart is opened.
- move to the abdomen and start squeezing the gallbladder to test patency.
- compress feces from rectum; place a ligature around the rectum and cut distal from the ligature; the entire intestinal tract can now be removed, using blunt and sharp dissection.
- remove both adrenals, and incise them with a scalpel.
- remove the kidneys by pulling the ureter caudally; peel capsules and cut to the pelvis and examine; the urinary bladder is opened at this stage.
- examine and remove the reproductive tract.
- remove the spleen from the stomach and incise.
- remove the skin and muscles from the head, and flex the head to locate the atlanto-occipital joint; cut to the ventral surface of the joint; cerebrospinal fluid can now be collected if necessary; cut the joint and neck muscles, and remove the head.
- remove the skull cap, and isolate the brain; examine the meninges; collect tissue for culture and histopathology.
- spread the intestines on a tabletop, and open the stomach, duodenum, sections of the jejunum, ileum, cecum, ascending colon, and descending colon; collect tissue if necessary.

Usually, the gastrointestinal tract is examined last in necropsies. If the clinical problem is related to the GI tract, it should be examined as early as possible, since it rapidly undergoes autolysis.

special issues regarding gene-modified swine

The ability to modify the mammalian genome by genetic engineering has allowed researchers to contemplate xenotransplantation from swine as a possible solution to the shortage of human donor organs, but gene-modified swine may also be used for creating new animal models.[130] Until recently, genetically engineering of swine was limited to microinjection of up to three foreign genes. In 2000 the first specific knockout pigs were produced, using a combination of gene targeting and embryo stem cell technique to knock out galacto-syltransferase, making xenotransplantation much more likely, although the efficiency still is lower than 1%.[131] Also in minipigs, transgenic techniques lead to a low efficiency.[132] The low efficiency of transgenic techniques may be overcome through cloning by somatic nucleus transfer, which makes it possible to create identical copies of founder transgenic pigs; but cloning still suffers from low efficiencies of around 3%.[133] Even when embryos do successfully implant in the uterus after nucleus transfer, pregnancies are often interrupted or animals die shortly after birth because of developmental abnormalities.[134] Cloned pigs have existed for too short a time to make a prediction on welfare effects of cloning. Genetic modification is often necessary for the purpose of immunomodulation during xenotransplantation, and pathological changes may be expected as a result of changed physiological traits.[135] Also, genetically induced diseases in swine models may have a serious impact on the animals' health, and special care is often necessary. However, studies with mice have shown that techniques used for gene modification as such are not necessarily associated with compromised animal welfare.[136] As with mice, developments in the field of gene modification of swine are incredibly fast, and it is expected that gene-modified swine will be increasingly used in biomedical research.[137,138]

resources

This chapter is meant to give an overview of laboratory-swine-related resources, without the aim of being exhaustive. It should be realized that resources, especially Internet resources, are time sensitive.

associations

AAALAC International
The Association for Assessment and Accreditation of Laboratory Animal Care International supports the use of animals to advance medicine and science when there are no non-animal alternatives, and when it is done in an ethical and humane way. When animals are used, AAALAC works with institutions and researchers to serve as a bridge between progress and animal well-being. This is done through a voluntary accreditation program in which research institutions demonstrate that they are not only meeting the minimums required by regulations, but are going the extra step to achieve and showcase excellence in animal care and use. AAALAC also has offices in Europe and Asia (see Internet address).

AAALAC International
5283 Corporate Drive, Suite 203, Frederick, MD 21703-2879
Phone: 301-696-9626—Fax: 301-696-9627
E-mail: accredit@aaalac.org—Internet: http://www.aaalac.org

AALAS
The American Association for Laboratory Animal Science is a forum for the exchange of information and expertise in the care and use

of laboratory animals. Since 1950, AALAS has been dedicated to the humane care and treatment of laboratory animals, and to the quality research that leads to scientific gains benefiting mankind and animals. AALAS has clinical veterinarians, technicians, technologists, educators, researchers, administrators, animal producers, and national and international experts as its members.

American Association for Laboratory Animal Science
9190 Crestwyn Hill Drive, Memphis, TN 38125
Phone: 901-754-8620—Fax: 901-753-0046
E-mail: info@aalas.org—Internet: http://www.aalas.org

AASV

The American Association of Swine Veterinarians is a nonprofit educational professional society organized to increase the knowledge of veterinarians in the field of swine medicine, elevate the standards of swine practice, promote the relationship between swine practice, the swine industry, and the public interest, promote the interests of swine veterinarians, improve the public stature of swine veterinarians, cooperate with veterinary and agricultural organizations and regulatory agencies, and promote goodwill among AASP members. AASP has approximately 2000 members hailing from practice, industry, and academia.

American Association of Swine Practitioners
902 1st Ave, Perry, IA 50220-1703
Phone: 515-465-5255—Fax: 515-465-3832
E-mail: aasv@aasv.org—Internet: http://www.aasv.org

ACLAM

The American College of Laboratory Animal Medicine is an organization of board-certified veterinary medical specialists who are experts in the humane, proper, and safe care and use of laboratory animals. ACLAM establishes standards of education, training, experience, and expertise necessary to become qualified as a specialist and recognizes that achievement through board certification. ACLAM actively promotes the advancement of knowledge in this field through professional continuing-education activities, the development of educational materials, and the conduct of research in laboratory animal medicine and science. ACLAM fosters the recognition of its members who contribute to human and animal health improvements by being the leaders of the veterinary medical specialty known as laboratory animal medicine.

The American College of Laboratory Animal Medicine does not have a permanent office address, but can be contacted through the president or secretary. Internet: http://www.aclam.org

ECLAM

The European College of Laboratory Animal Medicine (ECLAM) is a veterinary specialty organization established within the EC structure for veterinary specialization. The major tasks of the ECLAM are to improve prevention, diagnosis, control, and treatment of diseases of laboratory animals, prevention, alleviation, and minimization of laboratory animal pain and distress, and training of scientific, animal care, and ancillary staff.

ECLAM does not have a permanent office address, but can be contacted through the president or secretary. Internet: http://www.eclam.org.

ASLAP

The American Society of Laboratory Animal Practitioners promotes the acquisition and dissemination of knowledge, ideas, and information among veterinarians and veterinary students having an interest in laboratory animal practice. The Society does so for the benefit of laboratory animals, other animals, and society in general. ASLAP believes that laboratory animals do not differ from other animals in their need for proper preventive and therapeutic medical care, nutrition, physical, and psychological comfort, and all other elements that contribute to their health and well-being.

The American Society of Laboratory Animal Practitioners does not have a permanent office address, but can be contacted through the president or secretary. Internet: http://www.aslap.org

ESLAV

The European Society of Laboratory Animal Veterinarians gives veterinarians a forum to discuss issues that concern them in the field of laboratory animal science, in general and in Europe specifically.

The society's objectives are to promote and disseminate expert veterinary knowledge within the field of laboratory animal science. This is achieved through meetings, lectures, discussions, and publications. The advancement of veterinary knowledge and skills in subjects connected with the breeding, health, welfare, and use of laboratory animals is achieved by collaboration and exchange of information with other societies and allied scientific disciplines.

ESLAV does not have a permanent office address, but can be contacted through the president or secretary. Internet: http://www.aslap.org.

FELASA

The Federation of European Laboratory Animal Science Associations was founded in 1978. Its members are the European national laboratory animal science associations from Great Britain (LASA), Scandinavia (Scand-LAS), Germany (GV-Solas), France (SFEA), Spain (SECAL), Italy (CISAL), the Netherlands (NVP), Belgium (BCLAS), and Switzerland (SGV). As an international body, FELASA represents the common interests of the constituent associations by coordinating the development of education, animal welfare, health monitoring, and other aspects of laboratory animal science. Within the Council of Europe and the European Union, FELASA also has a political role, through offering advice to these bodies.

The Federation of European Laboratory Animal Science Associations holds office in the United Kingdom.

FELASA Secretariat
PO Box 3993
Tamworth, Staffs B78 3QU
United Kingdom
E-mail: felasaeu@felasa.eu—Internet: http://www.felasa.org

ICLAS

The International Council for Laboratory Animal Science was founded in 1956. Its aim is to promote and co-ordinate the development of laboratory animal science throughout the world, and promote international collaboration. One of its main activities is the harmonization of health monitoring programs, and to realize this, it has established reference health monitoring laboratories.

The International Council for Laboratory Animal Science does not have a permanent office address, but can be contacted through the president or secretary. Internet: http://www.iclas.org

ILAR

The Institute for Laboratory Animal Research is a division of the Commission on Life Sciences, one of eight major units within the National Research Council. The National Research Council is the working arm of the National Academy of Sciences, a private, non-governmental, non-profit organization chartered by the United States Congress in 1863 to "investigate, examine, experiment, and report upon any subject of science." ILAR works under the direction of a 15-member council. The council is composed of experts in laboratory animal medicine, zoology, genetics, medicine, ethics, and related bio-medical sciences. The council provides advice on all aspects of ILAR's program and formulates plans for the initiation of new programs.

Institute for Laboratory Animal Research
The National Academies
500 Fifth St. N.W., Washington, DC 20001
Phone: 202-334-2590—Fax: 202-334-1687
E-mail: ILAR@nas.edu—Internet: http://dels.nas.edu/ilar

National SPF Swine Accrediting Agency
The National SPF Swine Accrediting Agency, Inc., dates back to May of 1952. Its aim is to maintain and monitor hysterectomy-derived SPF swine colonies of accredited swine producers, to eliminate and prevent chronic growth-retarding diseases, and to enlarge the economic profit of swine producers. The National SPF Swine Accrediting Agency can be contacted to obtain addresses of accredited swine producers, in order to procure SPF swine for the laboratory.

National SPF Swine Accrediting Agency, Inc.
PO Box 280, Conrad, IA 50621
Phone: 641-366-2124—Fax: 641-366-2232
E-mail: spf@nationalspf.com—Internet: http://www.nationalspf.com

books

A range of reference books on laboratory swine are available. Three general books are:

Mount, L.E. and D.L. Ingram, *The pig as a laboratory animal.* Academic Press, London, 1971.

Pond, W.G. and K. Houpt, *Biology of the pig.* Comstock Publishing Associates, Ithaca, 1978.

Pond, W.G. and H. Mersmann, *Biology of the domestic pig.* Cornell University Press, Ithaca, 2001.

The symposium Swine in Biomedical Research has taken place three times thus far, and the proceedings were published as handbooks:

Bustad, L.K., R.O. McClellan and M.P. Burns, eds., *Swine in biomedical research.* Frayn & McClellan, Seattle, 1966.

Tumbleson, M.E., ed., *Swine in biomedical research,* vol. 1–3. Plenum Press, New York, 1986.

Tumbleson, M.E. and L.B. Schook, eds., *Advances in swine in biomedical research,* vol. 1–2. Plenum Press, New York, 1996.

One author, Michael Swindle, has published several handbooks on laboratory swine, anesthesia and surgery, and animal models:

Swindle, M.M., *Basic surgical exercises using swine.* Praeger Press, Philadelphia, 1983.

Swindle, M.M., *Swine as models in biomedical research.* Iowa State University Press, Ames, 1992.

Swindle, M.M., *Surgery, anesthesia and experimental techniques in swine.* Iowa State University Press, Ames, 1998.

Swindle, M.M., *Swine in the laboratory: Surgery, anesthesia, imaging, and experimental techniques.* CRC Press, Boca Raton, 2007.

A book on cardiovascular research using swine has been published. According to the authors, this book was compiled especially because "information on cardiovascular studies in swine is scattered widely throughout biomedical and agricultural literature, and few comprehensive reviews have been prepared":

Stanton, H.C. and H.J. Mersmann, eds., *Swine in cardiovascular research,* vol. 1–2. CRC Press, Boca Raton, 1986.

journals

The following relevant journals are placed in alphabetic order.

Comparative Medicine
Comparative Medicine (CM) is an international journal of comparative and experimental medicine, ranked by the Science Citation Index in

the upper third of all scientific journals. It advances knowledge about comparative medicine and laboratory animal science through the publication of scholarly articles about animal models, animal biology, laboratory animal medicine, laboratory animal pathology, animal behavior, animal biotechnology, and related topics. The journal publishes reports and reviews about basic and applied laboratory investigations, clinical investigations, and case studies, as well as informed and thoughtful opinions relevant to the humane care and use of laboratory animals. The journal is published six times a year.

American Association for Laboratory Animal Science
9190 Crestwyn Hill Drive, Memphis, TN 38125
Internet: http://www.aalas.org/publications/

ILAR Journal
ILAR Journal is the quarterly, peer-reviewed publication of the Institute for Laboratory Animal Research (ILAR), which is a unit of the National Research Council, National Academy of Sciences. *ILAR Journal* provides thoughtful and timely information for all those who use, care for, and oversee the use of laboratory animals. The audience of *ILAR Journal* includes investigators in biomedical and related research, institutional officials for research, veterinarians, and members of animal care and use committees.

Institute for Laboratory Animal Research
The National Academies
500 Fifth Street, NW, Washington, DC 20001
Internet: http://dels.nas.edu/ilar_n/ilarjournal/

JAALAS
The *Journal of the American Association for Laboratory Animal Science* (*JAALAS*) serves as an official communication vehicle for the American Association for Laboratory Animal Science (AALAS). The journal includes a section of refereed articles and a section of AALAS association news. In the refereed section of the journal, high-quality, peer-reviewed information on animal biology, technology, facility operations, management, and compliance as relevant to the AALAS membership is published. *JAALAS* publishes peer-reviewed research reports and scholarly reports. The journal is published six times a year.

American Association for Laboratory Animal Science
9190 Crestwyn Hill Drive, Memphis, TN 38125
Internet: http://www.aalas.org/publications/

Lab Animal

Lab Animal is a peer-reviewed journal for professionals in animal research, emphasizing proper management and care. Editorial features include new animal models of disease; breeds and breeding practices; lab animal care and nutrition; new research techniques; personnel and facility management; facility design; new lab equipment; education and training; diagnostic activities; clinical chemistry; toxicology; genetics; and embryology, as they relate to laboratory animal science.

Nature Publishing Group
75 Varick Street, 9th Floor, New York, NY 10013-1917
Internet: http://www.labanimal.com

Lab Animal Europe

Established in 2000 as the European version of the *Lab Animal* publication, *Lab Animal Europe* combines authoritative and highly respected editorial content for professionals and decision-makers in the laboratory animal research and biotechnology industries.

Agenda Marketing and Communication
PO Box 24, Hull, HU12 8YJ, United Kingdom
Internet: http://www.labanimaleurope.eu/

Laboratory Animals

Laboratory Animals is an international journal of laboratory animal science. *Laboratory Animals* is published by Laboratory Animals, Ltd., and is the official journal of the Federation of European Laboratory Animal Science Associations (FELASA).

The Royal Society of Medicine Press, Ltd.
1 Wimpole Street, London, W1M 8AE, United Kingdom
Internet: http://www.lal.org.uk

Pig International

Pig International is an international journal for swine producers on current worldwide topics and latest product and marketing news in the pig industry. It contains information on health topics, nutrition and breeding, and has many product resources. *Pig International* is published six times a year.

Watt Publishing, Co.
303 N. Main Street, Rockford, IL 61101
Internet http://www.piginternational-digital.com/piginternational/

internet resources

Many of the associations mentioned in the section on associations have links to other Internet resources relevant for laboratory animal science. A special type of resource is electronic databases for searching literature or scientific data. Only a small fraction of what is available is mentioned below.

Literature resources:

PubMed   http://www.ncbi.nlm.nih.gov/pubmed/

PubMed is a free service of the U.S. National Library of Medicine.

Web of Knowledge   http://www.webofknowledge.com/

The Web of Knowledge is published by Thomson Reuters and requires a subscription.

Scientific resources:

NAGRP Pig Genome Coordination Program, genome databases of the pig

http://www.animalgenome.org/pigs/

swine and diet, and equipment

Resources for swine, diet, and equipment are so diverse that a summary is not possible. The majority of swine used in research originate from agricultural sources, and local directories should be used. If SPF swine are required, the National SPF Swine Accrediting Agency can be contacted for local addresses. Miniature swine generally are supplied by specialized laboratory animal breeding companies. At least two buyer guides specialized in the laboratory animal field are available:

Lab Animal Buyers' Guide
Published yearly by *Lab Animal*: http://guide.labanimal.com/guide/index.html/

Laboratory Animals Buyers' Guide
Published every three years by Laboratory Animals Ltd.

Miniature Swine Resources

Miniature swine can be obtained from the following companies, among other sources.

Ellegaard Göttingen Minipigs (Göttingen minipigs)
Soroe Landevej 302, DK-4261 Dalmose, Denmark
Phone: (+45) 58185818—Fax: (+45) 58185880
E-mail: ellegaard@minipigs.dk—Internet: http://www.minipigs.com/

Marshall Farms (Göttingen minipigs)
Phone: 315-587-2295—Fax: 315-587-2109
E-mail: infous@marshallbio.com—Internet: http://www.marshallbio.com/

Sinclair Bio Resources (Sinclair minipigs, Yucatan mini- and micro-pigs, Hanford minipigs)
Phone: 573-387-4400—Fax: 573 387 4404
E-mail: info@sinclairbioresources.com—Internet: http://www.sinclair-bioresources.com/

Several other breeders offer miniature swine to the scientific community.

Diet Resources

Diet for miniature swine can be obtained from:

Bio-Serv
1 Eighth St., Frenchtown, NJ 08825
Phone: 908-996-2155—Fax: 908-996-4123
E-mail: sales@bio-serv.com—Internet: http://www.bio-serv.com/

Harlan Telkad
P.O. Box 44220, Madison, WI 53744-4220
Phone: 608-277-2070—Fax: 608-277-2066
E-mail: teklad@teklad.com—Internet: http://www.teklad.com/

Purina Mills
P.O. Box 66812, St. Louis, MO 63166-6812
Phone: 314-768-4592—Fax: 314-768-4859
E-mail: labdiet@purina-mills.com—Internet: http://www.labdiet.com/

Zeigler Bros
PO Box 95, Gardners, PA 17324
Phone: 717-677-6800—Fax: 717-677-6826
E-mail: info@zeiglerfeed.com—Internet: http://www.zeiglerfeed.com/

Special Diet Services
PO Box 705, Witham, Essex CM8 3AB, United Kingdom
Phone: (+44) 1376-511-260—Fax: (+44) 1376-511-247
E-mail: info@sdsdiets.com—Internet: http://www.sdsdiets.com/

Altromin
Im Seelenkamp 20, D-32791 Lage, Germany
Phone: (+49) 5232-608-80—Fax: (+49) 5232-608-820
E-mail: info@altromin.de—Internet: http://www.altromin.de/
Diets are produced by several other manufacturers.

references

1. Porter, V. 1993. *Pigs, a handbook of the breeds of the world.* Mountfield: Helm Information.

2. United States Department of Agriculture. 2008. *Agricultural Statistics 2008.* Washington: U.S. Government Printing Office.

3. Jones, G.F. 1998. Genetic aspects of domestication, common breeds and their origin. In *The genetics of the pig,* ed. M.F. Rothschild and A. Ruvinsky. Wallingford: CAB International.

4. Sambraus, H.H. 1992. *A colour atlas of livestock breeds.* London: Wolfe Publishing.

5. Ahamed, H.I., L.B. Becker, J.P. Ornato, et al. 1996. Utstein-style guidelines for uniform reporting of laboratory CPR research. *Circulation* 94:2324–36.

6. McGlone, J.J., C. Désaultés, P. Morméde and M. Heup. 1998. Genetics of behaviour. In *The genetics of the pig,* ed. M.F. Rothschild and A. Ruvinsky. Wallingford: CAB International.

7. Holtz, W. and P. Bollen. 1999. Pigs and minipigs. In *The UFAW handbook on the care and management of laboratory animals,* Vol. 1, Chapter 29, ed. T.B. Poole. Oxford: Blackwell Scientific.

8. Sippel, R.A. and B. Oldigs. 1982. Ethologischer Untersuchungen am Göttingen Miniaturschwein [Ethological study of the Göttingen minipig]. *Deutsche Tierärztliche Wochenschrift* 89(3): 97–113.

9. Sack, W.O. 1982. *Essentials of pig anatomy—Horowitz/Kramer atlas of musculoskeletal anatomy of the pig.* Ithaca: Veterinary Textbooks.

10. Dyce, K.M., W.O. Sack and C.J.G. Wensing. 1987. *Textbook of veterinary anatomy.* Philadelphia: Saunders.

11. Pond, W.G. and H. Mersmann. 2001. *Biology of the domestic pig.* Ithaca: Cornell University Press.

12. Mount, L.E. and D.L. Ingram. 1971. *The pig as a laboratory animal.* London: Academic Press.

13. Swindle, M.M. 1998. *Surgery, anesthesia and experimental techniques in swine.* Ames: Iowa State University Press.

14. Smith, A.C., F.G. Spinale and M.M. Swindle. 1990. Cardiac function and morphology of Hanford miniature swine and Yucatan miniature and micro swine. *Laboratory Animal Science* 40(1): 47–50.

15. Chowdhary, B.P. 1998. Cytogenetics and physical chromosome maps. In *The genetics of the pig,* ed. M.F. Rothschild and A. Ruvinski. Wallingford: CAB International.

16. Beynen, A.C., M.E. Coates and G.W. Meijer. 1993. Nutrition and experimental results. In *Principles of laboratory animal science,* ed. L.F.M van Zutphen, V. Baumans and A.C. Beynen. Amsterdam: Elsevier.

17. National Research Council. 1998. *Nutrient requirements of swine.* Washington: National Academy Press.

18. Schmidt, D.A. and M.E. Tumbleson. 1986. Swine hematology. In *Swine in biomedical research,* ed. M.E. Tumbleson. New York: Plenum Press.

19. Felsman, B.F., J.G. Zinkl and N.C. Jain. 2000. *Schalm's veterinary hematology.* Baltimore: Willey.

20. Rispat, G., M. Slaoui, D. Weber, P. Salemink, C. Berthoux and R. Shrivastava. 1993. Haematological and plasma biochemical values for healthy Yucatan micropigs. *Laboratory Animals* 27(4): 268–73.

21. Ellegaard, L., K.D. Jørgensen, S. Klastrup, A. Kornerup Hansen and O. Svendsen. 1995. Haematologic and clinical chemical values in 3 and 6 months old Göttingen minipigs. *Scandinavian Journal of Laboratory Animal Science* 22(3): 239–48.

22. Radin, M.J., M.G. Weiser and M.J. Fettman. 1986. Hematologic and serum biochemical values for Yucatan miniature swine. *Laboratory Animal Science* 36(4): 425–27.

23. Parsons, A.H. and R.E. Wells. 1986. Serum biochemistry of healthy Yucatan miniature pigs. *Laboratory Animal Science* 36(4): 428–30.

24. Tumbleson, M.E. and D.A. Schmidt.1986. Swine clinical chemistry. In *Swine in biomedical research*, ed. M.E. Tumbleson. New York: Plenum Press.

25. Hannon, J.P., C.A. Bossone and C.E. Wade. 1990. Normal physiological values for conscious pigs used in biomedical research. *Laboratory Animal Science* 40(3): 293–98.

26. Svendsen, P. and A.M. Carter. 1989. Blood gas tensions, acid-base status and cardiovascular function in miniature swine anaesthetized with halothane and methoxyflurane or intravenous metomidate hydrochloride. *Pharmacol. Toxicol.* 64(1): 88–93.

27. Cimini, C.M. and E. Zambraski. 1985. Non-invasive blood pressure measurement in Yucatan miniature swine using tail cuff sphygmomanometry. *Laboratory Animal Science* 35(4): 412–16.

28. Köhn, F., A.R. Sharifi and H. Simianer. 2007. Modeling the growth of the Goettingen minipig. *J. Anim. Sci.* 85(1): 84–92.

29. Köhn F., A.R. Sharifi, H. Täubert, S. Malovrh and H. Simianer. 2008. Breeding for low body weight in Goettingen minipigs. *Journal of Animal Breeding and Genetics* 125(1): 20-8.

30. Bouchard, G., R.M. McLaughlin, M.R. Ellersieck, G.F. Krause, C. Franklin and C.S. Reddy. 1995. Retrospective evaluation of production characteristics in Sinclair miniature swine—44 years later. *Laboratory Animal Science* 45(4): 408–14.

31. Panepinto, L.M. and R.W. Phillips. 1981. Genetic selection for small body size in Yucatan miniature pigs. *Laboratory Animal Science* 31(4): 403–04.

32. Johnston, N.A. and T. Nevalainen. 2003. Impact of the biotic and abiotic environment on animal experiments. In *Handbook of laboratory animal science*, ed. J. Hau and G.L. van Hoosier. Boca Raton: CRC Press.

33. Arrellano, P.E., C. Pijoana, L.D. Jacobson and B. Algers. 1992. Stereotyped behaviour, social interaction and suckling pattern of pigs housed in groups or in single crates. *Applied Animal Behavior Science* 35(2): 157–66.

34. National Research Council. 1996. *Guide for the care and use of laboratory animals*. Washington: National Academy Press.

35. Anon. 1986. European Communities Council Directive regarding the protection of animals used for experimental and other scientific procedures. *Directive 86/609/EEC*.

36. Georgiev, J., A. Georgieva, A. Kehrer and S. Weil. 1977. Beziehung zwischen Umgebungstemperatur, Luftfeuchtigkeit und Energie-umsatz beim Göttingen Miniaturschwein [Relation between environmental temperature, relative humidity and energy conversion of the Göttingen minipig]. *Berl. Münch. Tierarztl. Wochenschr.* 90(20): 392–96.

37. Tanida, H. and Y. Nagano. 1998. The ability of miniature pigs to discriminate between a stranger and their familiar handler. *Appl. Anim. Behavior. Sci.* 56(2–4): 149–59.

38. Degryse, A.D. and M.T. Calmettes. 1997. Effects of positive, minimal and adversive handling on the behaviour of pigs housed in "normal" or enriched conditions. In *Harmonization of laboratory animal husbandry*, proceedings of the sixth FELASA Symposium, ed. P.N. O'Donoghue. London: Royal Society of Medicine Press.

39. Gonyou, H.W., P.H. Hemsworth and J.L. Barnett. 1986. Effect of frequent interaction with humans on growing pigs. *Applied Animal Behaviour Science* 16(3): 269–78.

40. Ewan, R.C. 1991. Energy utilization in swine nutrition. In *Swine nutrition*, ed. E.W. Miller, D.E. Ullrey and A.J. Lewis. Boston: Butterworth-Heineman.

41. Ritskes-Hoitinga, J. and P. Bollen. 1997. Nutrition of (Göttingen) minipigs: facts, assumptions and mysteries. *Pharmacol. Toxicol.* 80 (Suppl. 2): 5–9.

42. Cuhna, T.J. 1966. Nutritional requirements of the pig. In *Swine in biomedical research*, ed. L.K. Bustad and M.P. Burns. Seattle: Frayn & McClellan.

43. Dettmers, A. 1968. On nutrition of the miniature pigs at the Hormel Institute. *Laboratory Animal Science* 18(1): 116–19.

44. Grieshop, C.M., D.E. Reese and C.M. Nyachoti. 2001. Nonstarch polysaccharides and oligosaccharides in swine nutrition. In *Swine nutrition*, ed. A.J. Lewis and L.L. Southern. Boca Raton: CRC Press.

45. Hart, R.W., D.A. Neumann and R.T. Robertson, eds. 1995. *Dietary restriction: Implications for the design and interpretation of toxicity and carcinogenity studies.* Washington: ILSI Press.

46. Thacker, P.A. 2001. Water in swine nutrition. In *Swine nutrition*, ed. A.J. Lewis and L.L. Southern. Boca Raton: CRC Press.

47. Hessler, J.R. and S.L. Leary. 2002. Design and management of animal facilities. In *Laboratory animal medicine*, ed. J.G. Fox, C. Lynn, F.M. Loew and F.W. Quimby. San Diego: Academic Press.

48. Russel, A.D., V.S. Yarnych and A.V. Koulikovskii, eds. 1984. *Guidelines on disinfection in animal husbandry for prevention and control of zoonotic diseases.* Geneva: World Health Organization (WHO).

49. Böhm, R. 1998. Disinfection and hygiene in the veterinary field and disinfection of animal houses and transport vehicles. *International Biodeterioration & Biodegradation* 41(3): 217–24.

50. Anon. 1996. *Manual for the transportation of live animals by road.* Redhill: Animal Transportation Association (ATA).

51. Anon. 1997. *Live animals regulations.* Montreal: International Air Transportation Association (IATA).

52. Bollen, P. and M. Ritskes-Hoitinga. 2004. The welfare of pigs and minipigs. In *The welfare of laboratory animals*, ed. E. Kaliste. Dordrecht: Kluwer Academic Publishers.

53. Friis, N.F. and A.A. Feenstra. 1994. Mycoplasma hyorhinis in the etiology of serositis among piglets. *Acta Veterinaria Scandinavica* 35(1): 93–98.

54. Scholl, T., J.K. Lunney, C.A. Mebus, E. Duffy and C.L.V. Martins. 1989. Virus-specific cellular blastogenesis and interleukin-2 production in swine after recovery from African swine fever. *American Journal of Veterinary Research* 50(10): 1781–86.

55. Hansen, A. K. 2002. Health Status and Health Monitoring. In *Handbook of Laboratory Animal Science*, Vol. 1, Chapter 11, ed. J. Hau and G.L. van Hoosier. Boca Raton: CRC Press.

56. Monshouwer, M., R.F. Witkamp, S.M. Nijmeijer, A. Pijpers, J.H.M. Verheijden and A.S.J. van Miert. 1995. Selective effects of a bacterial infection (*Actinobacillus pleuropneumoniae*) on the hepatic clearances of caffeine, antipyrine, paracetamol, and indocyanine green in the pig. *Xenobiotica* 25(5): 491–99.

57. Savlik, M., K. Fimanova, B. Szotakova, J. Lamka and L. Skalova. 2006. Modulation of porcine biotransformation enzymes by anthelmintic therapy with fenbendazole and flubendazole. *Research in Veterinary Science* 80(3): 267–74.

58. Krakowka, S., D.R. Morgan, W.G. Kraft and R.D. Leunk. 1987. Establishment of gastric campylobacter-pylori infection in the neonatal gnotobiotic piglet. *Infection and Immunity* 55(11): 2789–96.

59. Queiroz, D.M.M., G.A. Rocha, E.N. Mendes, A.P. Lage, A.C.T. Carvalho and A.J.A. Barbosa. 1990. A spiral microorganism in the stomach of pigs. *Veterinary Microbiology* 24(2): 199–204.

60. McNulty, M.S., G.M. Allan, G.R. Pearson, J.B. McFerran, W.L. Curran and R.M. McCracken. 1976. Reovirus-like agent (rotavirus) from lambs. *Infection and Immunity* 14(6): 1332–38.

61. Waldvogel, A.S., S. Broll, M. Rosskopf, M. Schwyzer and A. Pospischil. 1995. Diagnosis of fetal infection with porcine parvovirus by in situ hybridization. *Veterinary Microbiology* 47(3–4): 377–85.

62. Brunborg, I.M., C.M. Jonassen, T. Moldal, B. Bratberg, B. Lium, F. Koenen and J. Schonheit. 2007. Association of myocarditis with high viral load of porcine circovirus type 2 in several tissues in cases of fetal death and high mortality in piglets: A case study. *Journal of Veterinary Diagnostic Investigation* 19(4): 368–75.

63. Love, R.J., A.W. Philbey, P.D. Kirkland, A.D. Ross, R.J. Davis, C. Morrissey and P.W. Daniels. 2001. Reproductive disease and congenital malformations caused by Menangle virus in pigs. *Australian Veterinary Journal* 79(3): 192–98.

64. McTaggart, H.S., P. Imlah and K.W. Head. 1982. Causes of death and sex-differences in survival times of pigs with untreated hereditary lymphosarcoma (leukemia). *Journal of the National Cancer Institute* 68(2): 239–48.

65. Rottem, S. and M.F. Barile. 1993. Beware of mycoplasmas. *Trends in Biotechnology* 11(4): 143–51.

66. Kapikian, A.Z., M.F. Barilf, R.G. Wyatt, R.H. Yolken, J.G. Tully, H.B. Greenberg, A.R. Kalica and R.M. Chanock. 1979. Mycoplasma contamination in cell culture of Crohn's disease material. *Lancet* 2(8140): 466–67.

67. Ejsing-Duun M., J. Josephsen, B. Aasted, K. Buschard and A.K. Hansen. 2008. Dietary gluten reduces the number of intestinal regulatory T cells in mice. *Scandinavian Journal of Immunology* 67(6): 553–59.

68. Takahashi, T., T. Sawada, M. Muramatsu, Y. Tamura, T. Fujisawa, Y. Benno and T. Mitsuoka. 1987. Serotype, antimicrobial susceptibility, and pathogenicity of *Erysipelothrix rhusiopathiae* isolates from tonsils of apparently healthy slaughter pigs. *Journal of Clinical Microbiology* 25(3): 536–39.

69. Hansen, A.K., K. Dahl, D.B. Sorensen, E. Kemp and S. Kirkeby. 2004. Xenotransplantation: State of the art. *Acta Veterinaria Scandinavica* 99 (Suppl.): 7–12.

70. Brewer, L.A., H.C.M. Lwamba, M.P. Murtaugh, A.C. Palmenberg, C. Brown and M.K. Njenga. 2001. Porcine encephalomyocarditis virus persists in pig myocardium and infects human myocardial cells. *Journal of Virology* 75(23): 11621–29.

71. Hansen, A.K., H. Farlov and P. Bollen. 1997. Microbiological monitoring of laboratory pigs. *Laboratory Animals* 31(3): 193–200.

72. Patience, C., Y. Takeuchi and R.A. Weiss. 1997. Infection of human cells by an endogenous retrovirus of pigs. *Nature Medicine* 3(3): 282–86.

73. Martin, U. and G. Steinhoff. 1999. Porcine endogene Retroviren (PERV): In vitro Artefakt oder grosses Problem für die Xenotransplantation? [Porcine endogenous retroviruses (PERV): In vitro artifact or a big problem for xenotransplantation?] *Deutsche Tierarztliche Wochenschrift* 106(4): 146–49.

74. Cunningham, D.A., C. Herring, X.M. Fernandez-Suarez, A.J. Whittam, K. Paradis and G.A. Langford. 2001. Analysis of patients treated with living pig tissue for evidence of infection by porcine endogenous retroviruses. *Trends in Cardiovascular Medicine* 11(5): 190–96.

75. Herring, C., D.A. Cunningham, A.J. Whittam, X.M. Fernandez-Suarez and G.A. Langford. 2001. Monitoring xenotransplant recipients for infection by PERV. *Clinical Biochemistry* 34(1): 23–27.

76. Sangild, P.T., R.H. Siggers, M. Schmidt, J. Elnif, C.R. Bjornvad, T. Thymann, M.L. Grondahl, A.K. Hansen, S.K. Jensen, M. Boye, L. Moelbak, R.K. Buddington, B.R. Westrom, J.J. Holst and D.G. Burrin. 2006. Diet- and colonization-dependent intestinal dysfunction predisposes to necrotizing enterocolitis in preterm pigs. *Gastroenterology* 130(6): 1776–92.

77. Zhang, W., M.S.P. Azevedo, A.M. Gonzalez, L.J. Saif, T. van Nguyen, K. Wen, A.E. Yousef and L.J. Yuan. 2008. Influence of probiotic lactobacilli colonization on neonatal B cell responses in a gnotobiotic pig model of human rotavirus infection and disease. *Veterinary Immunology and Immunopathology* 122(1–2): 175–81.

78. Splichal, I., A. Splichalova and I. Trebichavsky. 2008. Cytokine response to *Escherichia coli* in gnotobiotic pigs. *Folia Microbiologica* 53(2): 161–64.

79. George, S., Y. Oh, S. Lindblom, S. Vilain, A.J.M. Rosa, A.H. Francis, V.S. Brozel and R.S. Kaushik. 2007. Lectin binding profile of the small intestine of five-week-old pigs in response to the use of chlortetracycline as a growth promotant and under gnotobiotic conditions. *Journal of Animal Science* 85(7): 1640–50.

80. Inman, C., R. Harley, G. Laycock, L. Sait, P. van Diemen, T. Humphrey, M. Stevens and M. Bailey. 2007. The effect of microbial colonisation on the development of the intestinal mucosal immune system in gnotobiotic pigs. *Immunology* 120 (Suppl. 1): 27.

81. Splichal, I., I. Rychlik, D. Gregorova, A. Sebkova, I. Trebichavsky, A. Splichalova, Y. Muneta and Y. Mori. 2007. Susceptibility of germ-free pigs to challenge with protease mutants of *Salmonella enterica* serovar Typhimurium. *Immunobiology.* 212(7): 577–82.

82. Travnicek, J., L. Mandel, I. Trebichavsky and M. Talafantova. 1989. Immunological state of adult germfree miniature Minnesota pigs. *Folia Microbiologica* 34(2): 157–64.

83. Stringfellow, D.A., K.P. Riddell and O. Zurovac. 1991. The potential of embryo transfer for infectious-disease control in livestock. *New Zealand Veterinary Journal* 39(1): 8–17.

84. Ellegaard, L. and A.K. Hansen. 1994. Produktion von mikrobiologisch definierten Miniatur-schweinen mit Rücksicht auf die Verbesserung des wissenschaftlichen Standards von Schweineversuchen [Production of microbiologically defined miniature pigs with the purpose of improving the scientific standard of research on swine]. *Der Tierschutzbeauftragte* 1:31–35.

85. Engstrand, L., S. Gustavsson, A. Jorgensen, A. Schwan and A. Scheynius. 1990. Inoculation of barrier-born pigs with *Helicobacter pylori*: A useful animal-model for gastritis type-B. *Infection and Immunity* 58(6): 1763–68.

86. Rehbinder, C., P. Baneux, D. Forbes, H. van Herck, W. Nicklas, Z. Rugaya, G. Winkler. 1998. FELASA recommendations for the health monitoring of breeding colonies and experimental units of cats, dogs and pigs: Report of the Federation of European Laboratory Animal Science Associations (FELASA) Working Group on animal health. *Laboratory Animals* 32(1): 1–17.

87. Hansen, A.K. 1993. Statistical aspects of health monitoring of laboratory animal colonies. *Scandinavian Journal of Laboratory Animal Science.* 20(1): 11–14.

88. Hansen, A.K. 1999. *Handbook of laboratory animal bacteriology.* Boca Raton: CRC Press.

89. Hansen, A.K. 1998. Microbiological quality of laboratory pigs. *Scandinavian Journal of Laboratory Animal Science.* 25 (Suppl. 1): 145–52.

90. Hansen, A.K. 1997. Health status of experimental pigs. *Pharmacology & Toxicology* 80 (Suppl. 2): 10-5.

91. Bollen, P.J.A. and L. Ellegaard. 1996. Developments in breeding Göttingen minipigs. In *Advances in Swine in Biomedical Research,* ed. M.E. Tumbleson and L.B. Schook. New York: Plenum Press.

92. NAGRP Pig Genome Coordination Program. 2009. http://www.animalgenome.org/pigs/nagrp.html/.

93. Howard, B., H. van Herck, J. Guillen, B. Bacon, R. Joffe and M. Ritskes-Hoitinga. 2004. Report of the FELASA Working Group on evaluation of quality systems for animal units. *Laboratory Animals* 38(2): 103–18.

94. Ingham, K.M., J.A. Goldberg, H.J. Klein, R.G. Johnson and M.D. Kastello. 2000. A novel approach for assessing the quality and effectiveness of IACUC oversight in investigator compliance. *Contemporary Topics in Laboratory Animal Science* 39(1): 28–31.

95. Taylor, D.J. 2006. *Pig diseases.* Glasgow: published by the author.

96. Straw, B.E., J.J. Zimmerman, S. D'Allaire and D.J. Taylor, eds. 2006. *Diseases of swine.* Oxford: Blackwell Publishing.

97. Smith, W., D.J. Taylor and R.H.C. Penny. 1989. *Wolfe colour atlas of pig diseases.* London: Wolfe Medical.

98. Bollen, P., O. Svendsen, K.D. Jørgensen, S. Klastrup and L. Ellegaard. 1997. Haematology of non-iron treated young Göttingen minipigs. In *Harmonization of laboratory animal husbandry,* proceedings of the sixth FELASA Symposium, ed. P.N. O'Donoghue. London: Royal Society of Medicine Press.

99. Svendsen, O., P. Bollen, K.D. Jørgensen, S. Klastrup and L.W. Madsen. 1998. Prevention of anaemia in young Göttingen minipigs after different dosages of colloid iron dextran. *Scandinavian Journal of Laboratory Animal Science* 25 (Suppl. 1): 191–96.

100. Madsen, L.W., A.L. Jensen and S. Larsen. 1998. Spontaneous lesions in clinically healthy microbiologically defined Göttingen minipigs, *Scandinavian Journal of Laboratory Animal Science* 25(3): 159–67.

101. Svensen, P. and C. Rasmussen. 1998. Anaesthesia of minipigs and basic surgical techniques. *Scandinavian Journal of Laboratory Animal Science* 25 (Suppl. 1): 31–44.

102. Flecknell, P. 1996. *Laboratory animal anaesthesia.* London: Academic Press.

103. Swindle, M.M. 2007. *Swine in the laboratory: Surgery, anesthesia, imaging, and experimental techniques.* Boca Raton: CRC Press.

104. Fish, R., P. Danneman, M. Brown and A. Karas. 2008. *Anesthesia and analgesia in laboratory animals.* San Diego: Elsevier.

105. Tranquilli, W.J., J.C. Thurmon and K.A. Grimm. 2007. *Lumb & Jones' veterinary anesthesia and analgesia.* Ames: Blackwell Publishing.

106. Wolfensohn, S. and M. Lloyd. 2003. *Handbook of laboratory animal management and welfare.* Oxford: Blackwell Publishing.

107. Mattinger, C., G. Petroianu, W. Baleck, W. Bergler and K. Hörmann, 2000. Nottracheotomie bei Göttinger miniaturschweinen [Emergency tracheotomy in Göttingen minipigs]. *Laryngo. Rhino. Otol.* 79(10): 595–98.

108. Kaiser, G.M., M.M. Heuer, N.R. Frühauf, C.A. Kühne and C.E. Broelsch. 2006. General handling and anesthesia for experimental surgery in pigs. *Journal of Surgical Research* 130(1): 73–79.

109. Becker, M. 1986. Anesthesia in Göttingen miniature swine used for experimental surgery. *Laboratory Animal Science* 36(4): 417–19.

110. Becker, M., R. Beglinger and H.A. Youssef. 1984. Isofluran beim Göttingen miniaturschwein [Isoflurane in Göttingen minipigs]. *Anaesthetist.* 33(8): 377-83.

111. Piermattei, D.L. and H. Swan. 1970. Techniques for general anesthesia in miniature pigs. *J. Surg. Res.* 10(12): 587–92.

112. Svendsen, P. 1997. Anaesthesia and basic experimental surgery of minipigs. *Pharmacol. Toxicol.* 80 (Suppl. 2): 23–26.

113. Boschert, K., P.A. Flecknell, R.T. Fosse, T. Framstad, M. Ganter, U. Sjøstrand, J. Stevens and J. Thurmon. 1996. Ketamine and its use in the pig. *Lab. Anim.* 30(3): 209–19.

114. Panepinto, L.M., R.W. Phillips, S. Norden, P.C. Pryor and R. Cox. 1983. A comfortable, minimum stress method of restraint for Yucatan miniature swine. *Laboratory Animal Science* 33(1): 95–97.

115. Noyes, E.P. and M. Fuentes. 1986. Restraint technique for intra-nasal inoculation and nasal culture of ten to twenty kilogram pigs. In *Swine in biomedical research*, ed. M.E. Tumbleson. Plenum Press, New York.

116. Baumans, V., R.G.M ten Berg, A.P.M.G. Bertens, H.J. Hackbarth and A. Timmermman. 1993. Experimental procedures. In *Principles of laboratory animal science*, ed. L.F.M van Zutphen, V. Baumans and A.C. Beynen. Amsterdam: Elsevier.

117. Brink, P. 2000. Continuous infusion in the minipig. In *Handbook of pre-clinical continuous intravenous infusion*, ed. G. Healing and D. Smith. London: Taylor and Francis.

118. Carroll, J.A., J.A. Daniel, D.H. Keisler and R.L. Matteri. 1999. Non-surgical catheterization of the jugular vein in young pigs. *Lab. Anim.* 33(2): 129–34.

119. Boogerd, W. and A.C.B. Peters. 1986. A simple method for obtaining cerebrospinal fluid from a pig model of *Herpes encephalitis*. *Laboratory Animal Science* 36(4): 386–88.

120. Faidley, T.D., S.T. Gallowa, C.M. Luhman, M.K. Foley and D.C. Beitz. 1991. A surgical model for studying biliary acid and cholesterol metabolism in swine. *Laboratory Animal Science* 41(5): 447–50.

121. Mortensen, J.T., P. Brinck and J. Lichtenberg. 1998. The minipig in dermal toxicology: a literature review. *Scandinavian Journal of Laboratory Animal Science* 25(suppl. 1): 77–84.

122. Vogel, B.E., M. Kolopp, T. Singer and A. Cordier. 1998. Dermal toxicity testing in minipigs: Assessment of skin reactions by noninvasive examination techniques. *Scandinavian Journal of Laboratory Animal Science* 25 (Suppl. 1): 117–20.

123. Bollen, P. and L. Ellegaard. 1997. The Göttingen minipig in pharmacology and toxicology. *Pharmacol. Toxicol.* 80 (suppl. 2): 3–4.

124. Gad, S.C., Z. Dincer, O. Svendsen and M. Tinglef Skaanild. 2007. Minipigs. In *Animal models in toxicology*, ed. S.C. Gad. Boca Raton: CRC Pess.

125. Jørgensen, K.D., T.S.A. Kledal, O. Svendsen and N.E. Skakkebæk. 1998. The Göttingen minipig as a model for studying the effects on male fertility. *Scandinavian Journal of Laboratory Animal Science* 25 (Suppl. 1): 161–70.

126. Jørgensen, K.D. 1998. Teratogenic activity of tretinoin in the Göttingen minipig. *Scandinavian Journal of Laboratory Animal Science* 25 (Suppl. 1): 235–43.

127. Anon. *OECD guidelines for the testing of chemicals.* http://www.oecd.org/.

128. Anon. *OECD guidelines for the testing of chemicals,* Section 4: health effects. 1981. Subchronic oral toxicity, non-rodent: 90 days, Guideline 409, updated 1997.

129. Bailie, M.B., S.K. Wixson and M.S. Landi. 1986. Vascular-access-port implantation for serial blood sampling in conscious swine. *Laboratory Animal Science* 36(4): 431–33.

130. Du, Y., P.M. Kragh, Y. Zhang, J. Li, M. Schmidt, I.B. Bøgh, X. Zhang, S. Purup, A.L. Jørgensen, A.M. Pedersen, K. Villemoes, H. Yang, L. Bolund and G. Vajta. 2007. Piglets born from handmade cloning, an innovative cloning method without micromanipulation. *Theriogenology.* 68(8): 1104–10.

131. Nottle, M.B., A.J.F. d'Aprice, P.J. Cowan, A.C. Boquest, S.J. Harrison, C.G. Grupen. 2002. Transgenic perspectives in xenotransplantation. *Xenotransplantation.* 9(5): 305–08.

132. Uchida, M., Y. Shimatsu, K. Onoe, N. Matsuyama, R. Niki, J.E. Ikeda and H. Imai. 2001. Production of transgenic miniature pigs by pronuclear microinjection. *Transgenic Res.* 10(6): 577–82.

133. Giles, J. and J. Knight. 2003. Dolly's death leaves researchers woolly on clone ageing issue. *Nature.* 421(6925): 776.

134. Pennisi, E. and Vogel G. 2000. Clones, a hard act to follow. *Science* 288(5475): 1722–27.

135. Olsson, K. 2000. Xenotransplantation and animal welfare. *Transplantation Proceedings* 32:1172–73.

136. Van der Meer, M. 2001. Transgenesis and animal welfare. Thesis, Department of Laboratory Animal Science, Utrecht University.

137. Prather, R.S., R.J. Hawley, D.B. Carter, L. Laia and J.L. Greenstein. 2003. Transgenic swine for biomedicine and agriculture. *Theriogenology.* 59(1): 115–23.

138. Sachs, D.H. and C. Galli. 2009. Genetic manipulation in pigs. *Current Opinion in Organ Transplantation* 14(2): 148–53.

index